THE TIMEKEEPERS
OF
ANCIENT EARTH

The stone structures and megalithic arrangements
preserve the evidence of the technological
achievements of an advanced civilization.

Arturo Villamarin

Book cover design and graphics by the author, except as noted.

TO MY FATHER

My father was a self-made civil engineer, agronomist and surveyor. From an early age he taught my older brother and me how to assist him as his chainmen; to take the measurements for topographic mapping for his work. He taught us how to use the optical transit instrument and how to draw the topographic maps from the data we collected.
We learned his work ethic and sense of duty and above all, his love of all knowledge. Of particular interest to him was history, which he could recollect in detail and with it enthrall an audience. An audience that was always there; for real history or tall tales, of which he was a master!

PREFACE

The archaeological record presents a unique opportunity for the study of its physical remains in terms of architectural layout; the site's configuration and the monuments' azimuth, without being concerned for a moment about the; who, when and how these were created. The physical objects at their sites contain intrinsic measurable parameters which follow basic mathematical principles. Most of the sites' objects can be evaluated; in their design, size and layout, in addition to their global location and positioning at the site; all of these parameters can be instrumentally ascertained with enough accuracy.

The impetus for our evaluation in those terms was to address the **'why'**. *What justifiable need made the architects of the archaeological record invest the time, manpower, effort and economic cost spent in planning and building, even the smallest menhir alignment or circle? The simple act of standing a menhir or megalith stone implies an expenditure of dearly available resources. Can we glean what that need was, can it be found in their measurements? Was there a greater benefit in their design beyond indicating the solstices and equinoxes for agricultural or religious purposes as it is believed by many researchers? This last question is the engine that has moved this line of inquiry forward. This study was focused on the analysis of each site by taking measurements of each structure and or the layout of the site to answer those questions and others such as: Are their sizes, heights and shapes significant? Are these monument sites geographically related to others around the world in any measurable way?*

The results of the various studies included in this work show that the design geometry and location of every structure had a deliberate purpose. The significance of each structure be it a glyph, a megalith, an individual structure or an arrangement of structures at a site, regardless of size or whether they were inhabited or not, most of them encode the direction to other places or other structures.

Several alignments between structures were found to be reciprocal and in a few instances, at a single site, the alignment of small structures within it point in the direction of other sites thousands of miles away.

The alignment of structures, in itself, does not justify the need and effort to build them, unless an underlying reason is identified. In the study we uncovered that the relative location of many of the sites encode astronomical facts about earth such as the presession of its orbit. Also, the physical measurements of some of the structures reveal their designs encode other astronomical properties of earth, such as its axial tilt and confirm their alignment with the solstices. The locations and design of a few archaeological sites reveal their designers knew of the existence of the earth's ridge lines.

AUTHOR'S NOTE

This book, we suggest be read following the discussion in Google©
earth or a similar program. All the data files used in this research
are available, to obtain a list contact: villarts@charter.net. The
geographical coordinates for most archaeological sites mentioned
are given within the text. You are encouraged to do so, and while
in Google© earth, avail yourself of the photographs posted by its
users. The experience will be rewarding for anyone interested in
archaeology and geography at any level, or to those who love to
travel.

The methodology used for this research is explained in detail in
Chapter 14. It is easy to follow and the reader is encouraged to
use this methodology to confirm our findings.

The longitude and latitude coordinates are given in decimal
degrees. You may set Google© earth to display readings in decimal
degrees by choosing this option under Tools-Options. Although
there is a one hundredth of a degree implied on the
measurements that are given, the error in a given measurement
will depend on several factors; distance, direction and starting
point of a measurement, which can introduce the largest error in
measurement. To reproduce a measurement one must zoom-in
to start at the same point. For the study we assumed 1° error. GE
historical mapping does not match locations perfectly every time,
a 'drift' is observed occasionally.

In this work we use the local spelling of places' names whenever
possible.

AKNOWLEDGMENTS

I'd like to acknowledge Ron and Melanie Vandervalk, whose care and support, allowed me to focus on the research and completion of this work.
I'd also like to acknowledge my wife for her assistance editing and my sixth grade world geography teacher, Mr. Nieto. He taught geography by having everyone draw the maps in excruciating detail. To this day I still remember how to do it. Maybe not accurately, but the cartographic ability to read them remains.

Arturo Villamarin

INDEX

S I O- GLYPH

NORTH

CANDELABRA

CHACO GLYPH

S P O- GLYPH

ANTENNA GLYPH

MONKEY GLYPH

MOAI

PROLOGUE

EARTH's ARCHEOLOGICAL WONDERS was there a purpose built into their designs?

Understanding our past can become an obsession. Every stone, every bone and every vestige of man's presence, in the precise context, captures our imaginations. We need to know. It is, or it seems to be, encoded in our genes. Man's drive to acquire knowledge, especially about him is so strong that more often than we would like to admit, it drives us to conjure up fantasies and myths as explanations for those things that are not patently obvious. For primitive peoples this was the actual process used to begin understanding life, at least until patterns began to emerge from every day phenomena. Science, or the scientific approach to understanding these phenomena, sought to capture perceived reality and reduce it to reproducible observations and results. Reproducibility of results means, as defined by IUPAC: "Reproducibility is the ability of an entire **experiment** or study to be reproduced, either by the researcher or by someone else working independently. It is one of the main principles of the scientific method and relies on *ceteribus paribus*-all being equal"[2].

It also means that a newcomer who can alter or improve the observation, must follow the given constraints or show how the proposed new constraint or precision provides a better, more universal explanation that improves the process, or in itself advances understanding. Archaeology is, in particular, a science that seems to require a vision that maintains a delicate balance between intuition and scientific accuracy. The head and heart, the emotional and factual need to be brought to bear on the evaluation or assessment of the unknown encoded in the vestiges of the past. The origin of man is perhaps the most debated topic, advanced through many disciplines and theories, even those classified as 'fringe'. One of those fringe theories is the visitation of earth by extra terrestrials in antiquity. This is a theory that will probably outlive mankind itself, without a proof. It proposes a plausible pathway that could explain how the human race became what it is today. Another theory speaks of advanced earlier civilizations; both rely on our inability to explain how megalithic structures were built thousands of years ago presumably without technology. This research focuses on the physical measurement of objects which reflect the technological knowledge that was used to create them.

We start by making reference to existing theories and recall them when the physical data highlights some of their aspects or elements; that is, we used them as a backdrop by testing the claims they make which appear to be relevant to the study.

CHAPTER 1

INTRODUCTION TO THE LEGACY

 In 1967, Swiss author E. von Däniken became one of the foremost proponents of alien beings visitations of earth. He proposed the theory, in his book "Chariot of the Gods". Intuitively he believed the Nasca lines were the proof of his theory; however, he wasn't able to demonstrate what his gut was telling him. Since the publication of his book his theories were dismissed by the scientific community as 'quackery'. Still, today his writings appear with others of this kind on shelves marked, from mysticism, to at best, speculation.

Perhaps his lack of scientific education in the early days contributed to this off-hand rejection. Today his following has grown in popularity and greater acceptance.

Author Zacheria Sitchin (1920-2010) claimed he was, "One of the few scholars able to read and interpret ancient Sumerian and Akkadian clay tablets. Sitchin based his bestselling *The 12th Planet* on texts from the ancient civilizations of the Near East". Sitchin.com

What he deciphered from these texts corroborated some of Von Däniken's ideas, and provided more detailed analysis in his book series, "The Earth Chronicles", not without garnering skepticism of its own.   In both their accounts, ancient peoples witnessed the arrival of extra terrestrials who left their culture in their hands. Sumerians, Egyptians, Aztecs, Incas and Mayas and others recorded the visitors' appearance as powerful beings or angels. The visits were documented in their scriptures, as well as, in monuments, statues, carvings, gold artifacts and glyphs in various forms, we now find in the above societies' and many other's archaeological records around the world. Besides the generally accepted belief that this legacy came to us as the product of our current civilization there are other theories; the existence of a much earlier technologically advanced civilization on the planet, is one that has gained a following in recent years. And in some instances use the biblical recounts of cataclysmic destruction as the reason for their disappearance.

THE LEGACY

ARTIFACTS
Some of the physical evidence without a definitive explanation seems to add weight on the side of 'proof' of the visits by extra-terrestrial beings. For example, in the Gold Museum in Bogotá, Colombia, there are multiple gold figurines in the shape of 'delta wing aircraft'. A claim vehemently denied by historians, down to the tour guides at the site. There are also fully armored figures, some angel like! As shown below. The motifs these have can also be found in many places the world over, as megaliths, artifacts or glyphs.

Photos by the author

PLACES

In Tambo, Perú there is a geo glyph site known as the Band of Holes, another archaeological wonder. This band consists of man-sized holes forming a necklace up the mountain to an altitude of about 600 feet from the valley, shown in the graphic below as small circles in gray. Atop this band, immediate to the left there is an oval shaped valley, which has curious striations. The striations fan out downwards forming a 'flowing skirt', while at the top of the striations, where they start; there is an arc facing upwards. On the right side, the arc continues over the mountain ridge ending on a straight line pointing south east. A similar line is found at the other end of the arc, but, it points south west. With a little imagination we make out the overall figure in the valley might look like an 'Angel' (13.7°S 75.87°W) as shown on the graphic below. This 'Angel' we find is curiously reminiscent of the golden 'Angel' shown above. It is remarkable that these two works were produced by two civilizations some 1300 miles apart. We found no documentation for the 'Angel' figure in Tambo Perú. Similar arrangements of 'holes' are found in diverse places on earth in the form of cups carved on stones. In southeast US; the Judaculla rock and the Forsyth stone, the Sonora stone in Mexico and the Duddington rocks in Ireland and the Shere Hills band in Nigeria, Africa and Göbekli Tepe, to mention a few.

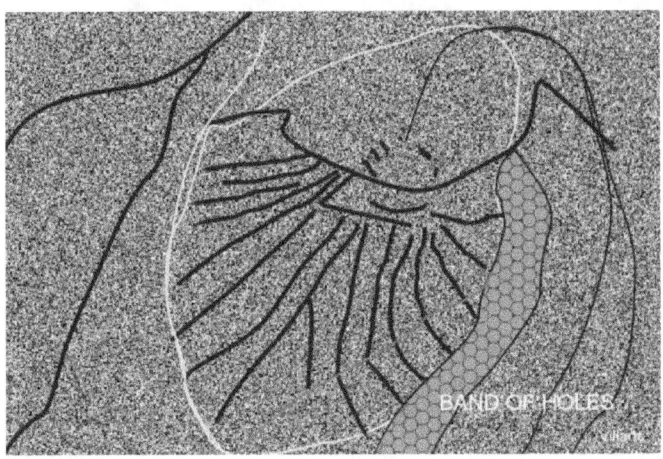

BAND OF HOLES

Another type of historical site of interest is the mound. Mounds can be counted in the thousands found around the world. Mounds can have many shapes; in Ohio, US; the Newark rings and the famous Serpent (39.03N 83.88W) or, in Louisiana US, where several types are located; we find a round mound and an amphitheater -like configuration together, known as Poverty Point (32.64°N, 91.41°W). This is the site of an early civilization in the lower Mississippi Valley (2200BCE-700BCE). Others pyramidal found in Mexico; Iguatziu and Cerrito and circular in Denmark known as Trelleborgs, which are believed were Viking fortresses.

Wikimedia Commons Image

The Poverty Point site is separated in time and distance from another amphitheater, similar to it near Cusco Perú where the Inca civilization prospered (300BC-1700AD).

Although similar in design, is much grander in scale; there are three of these at the same location, although the other two are smaller.

Moray Inca- Wikimedia Commons Image

The site is called Moray Inca. It is located at (13.32°S 72.18°W), some 3,408 miles away. This amphitheater design is not unique to these locations. There are several other examples throughout the Inca civilization and other locations around the world.

SYMBOLS

THE LEMNISCATE
There are other sites of greater interest from the technological point of view. At these sites, glyphs are found that have implicit mathematical meaning; other than triangles, squares or circles. Their geometrical meaning in terms of space flight makes them significant for the proponents of alien visitations.

One of these glyphs represents the sideways number eight commonly known as the infinity sign. This symbol has been known since the 5th century AD and is attributed to Proclus, a Greek mathematician.

Its use in mathematics is credited to mathematician John Wallis, who used it first in 1655, to mean infinity. Eight-shaped curves of different configurations were later derived by several mathematicians: Booth, Bernoulli, Huygens and Viviani. Each one of these different 'eight-shaped' curves corresponds to a different polynomial equation[22].

THE ANALEMMA

Another one of these curves is known as an Analemma. An Analemma curve is the path traced by a heavenly body onto another, such as the sun onto the earth, as the position of the sun changes in the sky with each rotation of the earth. This phenomenon also occurs, planet to planet and geosynchronous satellites and earth.

Satellite reception devices need to track their specific trajectory to optimize the signal reception, so they are programmed to track the satellite's Analemma curve. The picture below shows the various positions of the sun, as it appears in the sky, photographed from the same location.

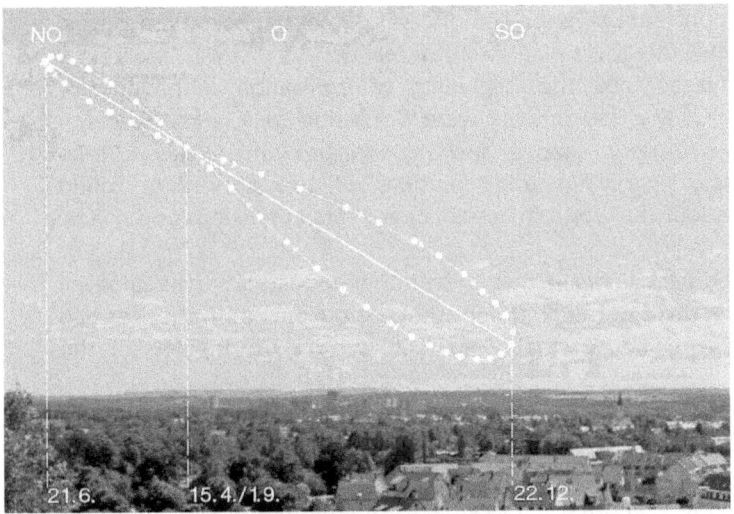

Analemmas or eight-shaped glyphs are found in different forms and at many locations. Some of these curves are found at archaeological sites located remotely from each other. For example, The Sonora Stones, in Sonora, México (27.83°N, 109.48°W) and in Nasca, at least at two locations 14.684° S 75.113°W and 14.79°S 75.28°W, the latter is the largest we have found, it is 475 ft. long. These two sites are 3,721 miles apart. Three other lemniscates were found at Kirbeth Al Umbashi in Jordan: 33.050°N 36.995°W - 32.882°N 36.953°W - 32.818°N 36.927°W, about 8,000 miles from Nasca.

The symbol is, also, found in Easter Island not as a glyph but as a structure. This stone structure is called a Manavai. The cross section of the Manavai structure, as viewed from above, is an eight-shaped curve.

One of these structures is located near Ahu Tepeu at (27.1°S 109.41°W) 3,776 miles directly south of Sonora, and 2,365 miles south west from Nasca. The shape of this stone structure is like a Bernoulli's Lemniscate.

The importance given to Lemniscates curves in ancient times was, probably, due to its mathematical reality. This fact was apparently recognized since the beginning of civilization, although not explicitly. Mystical powers were attributed to it, as it was with other geometric shapes, such as triangles and circles. Stone carvings, engravings and paintings of this curve are found throughout the world; this symbol appears in religious and mystic art, as well.

It could be significant that this symbol appears at archaeological sites, in favor of the argument that the builders of these sites had flight capability, or that alien visitors were the builders of the sites.

THE SPIRAL

The Spiral is another curve which is mathematically defined; it is found at many locations around the world.

Wikimedia Commons Images- Newgrange

Its mathematical form was first defined by the Greek scientist and Mathematician Archimedes (287-212 BC) and today is known in mathematics as the Archimedean Spiral. One obvious association of spirals with space travel is with the spiral galaxies.

At Verneukpan, South Africa (30.0°S 21.06°E) in an area in the desert of approximately sixteen square miles, there are one hundred or so spirals scrapped on to the ground and probably as many are also found in the Nasca plain in Perú, including the tail of the Monkey geo glyph. At Nasca, the most prominent spiral is found at (14.68°S, 75.36°W) and the Monkey is found at (14.71°S 75.14°W). The distance between the location of Verneukpan and Nasca is 6,086 miles. Spirals are, also, found at various other archaeological sites as megalithic art, places such as Newgrange and Chang'an -Xian, China[22]. The Newgrange spiral decorations carved on a monolith at the entrance of the mound is shown above. Also, there are three spirals on the back of Moai RR-001-156 in Easter Island, which was unearthed by EISP-Easter Island Statue Project.

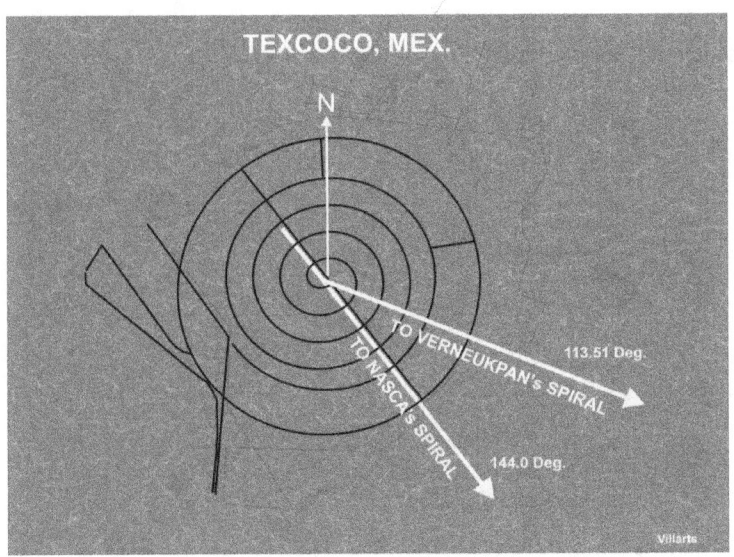

The largest spiral known is about 1.8 miles in diameter; it is found in Texcoco, México (19.57ºN 99.0ºW), shown above. This spiral has an obscure history; it is located in an area that was part of a lake in pre-Columbian times, now mostly drained as is most of Tenochtitlán, today Mexico City.

Some believe it was constructed in the 50's by a salt recovery company to use for desiccation. It appears that its archaeological value was cloaked behind this account, which otherwise could have prevented the exploitation of the salts from the lake. Furthermore, there is no technical reason salt desiccation would require this design.

What we found about this site and its design makes a strong case for it to be classified as an archaeological treasure. The Texcoco spiral, unlike any known others, its axis and origin are part of the design. Following its axis line southward extended as a great circle, the line connects with the main spiral at Nasca. The Nasca spiral is a Fermat spiral, the same as the designs at Newgrange, while the one at Texcoco is an Archimedean spiral, as are the Verneukpan spirals, to which Texcoco's is also connected. The arrangement and significance of the Verneukpan spirals will be discussed later in chapter 8.

Wikimedia Commons Image

There may, also, be a connection between space travel and spirals. The Euler spiral -similar to the Monkey's tail- is known as a transition spiral. It changes linear direction into circular direction. The picture above is the Esperanza Stone found in México in 1909. The arrow shows the double spiral. The picture below is an Euler's double spiral. In Blythe, California (33.8°N 33.53°W), a double spiral appears below the horse glyph.

Half of an Euler's spiral would be similar to the glide trajectory a vehicle from outer space would have during entry into the earth's atmosphere.

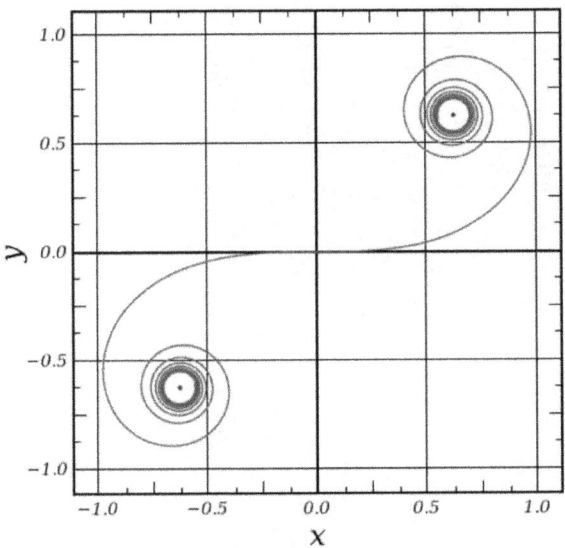

Wikimedia Commons Image.

A glide trajectory (with an atmosphere entry angle of about 6°) is necessary to be able to reuse the vehicle; such as with the NASA space shuttle[33]. Coincidentally this is the shape of the Monkey's tail.

GLYPHS

There are, literally, thousands of glyph types, sometimes hundreds are found on a single rock. The Esperanza stone shown above is an example as are Sonora and the Judaculla Stones, these two will be discussed later.

Amongst the many glyphs there are ones that repeat in so many places that after awhile they begin to seem commonplace.

The double spiral is one of them, as is the 'infinity' symbol. These are fairly easy to explain, as we have attempted above. Others boggle the mind as to what they might mean.

NASCA 'HANDS' GLYPH

Villarts

The majority of glyphs fall under this category and we just dismiss them as 'interesting' and don't give them a second thought, unless you happen to be a professional epigrapher (glyph reader) or their characteristics repeat enough times that the archaeology enthusiast can, after viewing a few, begin to recognize patterns.

One such instance is a curious glyph found on the Nasca plateau, known as "Hands", shown above. Indeed, they do look like hands; except, one hand has only four fingers and they are attached to what appears to be a bird. The Monkey glyph has the same finger count. This is interesting in itself, but, 4,098 miles away in the Chaco canyon in New Mexico in a cave, known as Una Vida- One Life- there is a glyph we call Chaco -shown below. This glyph is a figure with antennae which on his right hand holds a spiral and his left displays four fingers. There are at least two others in the cave with their hands spread out displaying four and five fingers, as well. The Orongo petro glyph, 4,351 miles away at Easter Island, also, shows a figure with four and five fingers.

Easter Island, Orongo and Chaco canyon Glyph -Wikimedia

The 4-5 finger arrangement is also displayed in the Palma Sola petro glyphs in México and the glyphs at Chelly Canyon in Arizona and the Newspaper glyphs in Utah, US.

CIRCLES

Circular stone arrangements are found in many designs, degree of structural sophistication and sizes; as plain circles made up of small stones or uncut megaliths weighing up to several tons, some arranged with center stones and spokes like cart wheels.

The largest most sophisticated arrangement made of carved megaliths is Stonehenge.   Some of these circles are named; Wheels, Calendars or Sun Dials and observatories. They are an archaeological record of a different kind and of special interest. These stone circles are found in many locations the world over, particularly, in the British Isles, and have been the subject of intense   study.   The   science   of,   Astroarchaeology   or Archeoastronomy came about from the results of many studies some which showed a connection exists between these ancient megalithic circular structures and stars, planets and the earth's solstices. We'll explore these concepts later in the study.

Medicine Wheel-US Forest Service photograph.

This science, also, studies the alignment of other archaeological structures, not only those that are similar in design to Stonehenge, but also, Pyramids and Temples such as; Giza, El Castillo at Chichén Itza, Kalsasaya in Tiwanaku and others.

The alignment some of these structures have with the solar solstices, planets and constellations are reported in the literature. Those studies have yielded hypotheses or understandings regarding the degree to which ancient cultures understood and might have used the information for agricultural and or ceremonial purposes. However, they do not explain or justify the economic value of the enterprise for the larger monuments. More significant is the fact that agriculture flourished around the world without the aid of these structures; which make that hypothesis more romantic than factual. The connection with the sun cycle can be, and will be evaluated as a physical reality, from an astronomical perspective in chapter 6. Also, there is no science that explains why the designs and geometries of many of the structures are similar. Attempts have been made to correlate or find out if there is a relationship between the structure's locations relative to each other, or their geographical positioning without yielding verifiable certainty. An interesting analysis is provided by Jim Alison[12], who claimed there is a correlation between the location of archaeological sites and the North Pole. In the end, the theories don't tell us why they were built. Regarding megalithic circles, whether they are found in Europe, the Americas, Israel or India, they appear to have similar layouts. The basic issues outlined above have driven this research; the results of which provide a new perspective.

The picture on page 26 is; the Medicine Wheel found at Big Horn Wyoming and the other below is 7,668 miles away at Junapani, India. Other circles similar to these are found at Hanga Te'e Easter Island-Chile, Nasca-Perú, Monsaraz and Almendres in Portugal, Nabta Playa-Egypt, several others in the UK besides Avebury and Stonehenge and the Golan circle in Israel and the Trelleborg rings in Denmark and the sites of Sine and Junapani which contain about thirty circles each; to name a few. The Medicine Wheel, in particular, appears to have played an important geographic role; as we'll see its location will be visited time and again throughout this research.

Junapani,India -Ganesh Dhamodkar/Wikimedia Commons

Comparisons, have been made for other structure types, such as: pyramids, dolmens and Temples, because of the similarities they share in their architecture and engineering, in spite their being found in many diverse places on earth. The similarities in design and construction these structures share, conjure up an enigma many scientists have questioned: how is it possible that cultures separated by thousands of miles, managed to design and engineer virtually identical structures and in some instances, apparently used similar building techniques. Pyramids, in particular, are examples of this. They are found in Mexico, China, Egypt, Peru and many other places. And the lore and facts regarding them abounds.

The architectural model of the Pyramid at Cholula México, displayed below, is of one of the largest pyramidal structures in the world, not in height, but at the base. It is only about half the height of the Chephren pyramid, but, at its base is twice as wide.

Model Pyramid of Cholula, México- - Wikipedia Commons

This pyramid, is believed, was built over a thousand years starting in about 300BCE. The pyramid was enlarged layer over layer, becoming taller each time. The last modification was done by the Spanish Conquistadores in the name of the Church. Still today, it displays a cathedral on top of it, a symbol of the intent the Church had to squelch local traditions. That onslaught, not only philosophical but physical, together with diseases eventually decimated the local population.

This layering of construction is not unique, archaeologists using the latest radar techniques have found previously unknown under-laying construction at some of the places mentioned above.

MEGALITHIC STRUCTURES

Gate of the Sun Tihuanaco, Bolivia- Wikimedia Commons Images

These structures, as a whole, range in engineering design sophistication, from the simple dolmen; a structure consisting of two pillars and a cap stone known as lintel, to the Egyptian Temples and pyramids.

Megalithic structures are found in places such as; Tihuanaco, Bolivia and as far as the pacific island of Tonga. Megaliths are, also, found around the globe by themselves, as Cairns -stacked stones such as Inukshuks, or as part of Temples with awe inspiring engineering, architectural achievement and artistic beauty in astonishing detail; a salient example is the temple of Luxor in Egypt. Building these large structures required mathematical precision in their measurements; during design and the stone cutting and alignment, which had to be orders of magnitude more stringent compared to that for the construction of dolmens or stone circles. These achievements are hard to explain considering the low level of technical advancement at the time when they were built. However, the technology required for stone circles and stone menhir alignments, Cairns and Ahus and their execution have been explained, in some instances as the product of human awareness, inquisitiveness and penchant for discovery, which is displayed in their construction and in their alignment following the trajectories of stars, planets, the sun or the moon.

El Infiernito, Villa de Leyva -Colombia- Wikimedia Commons

A typical monolith menhir alignment is shown above. The El Infirnito at Villa de Leyva, Colombia is a menhir alignment at an angle of 90°E, located about 4° degrees north of the Equator at an elevation of about 7,000 feet above sea level. Alignments, such as this one are not only related to astronomical events, but in some instances are part of a grander design, as this research has reveled.

The importance of the geographical location and azimuth alignment of structures within a site like this will be visited and explained in detail later.

TECHNOLOGY
The technologies that must have been used to build the pyramids, and were required for their execution, were unknown until about 800 BC when algebra was first developed in Damascus. Algebra and geometry advanced, not until Pythagoras and Euclid formally defined geometry and space (580 to 265 BC). These sciences were and are, necessary for basic construction, as is Newtonian mechanics. Mechanics is the science needed to set up the scaffolding required to move megaliths; some of them weighing up to hundreds of tons. A sample is shown below.

Sir Isaac Newton (25 December 1642 – 20 March 1727) did not publish the law of mechanics until 1687, in his *Philosophiæ Naturalis Principia Mathematica*. All of the major structures being considered are believed to precede mathematics by hundreds, if not thousands of years.

Baalbek Thrilithon- Wikimedia Commons Images

There are other factors that we think contribute to the study and evaluation of how these projects where carried out by our current civilization. One of these factors is logistics; the project management and control skills that were necessary and that had to be available prior to the initiation of anyone of the major works mentioned. These skills were required not only for the daily planning, but, through the years it took to finish anyone of those projects. How these projects were accomplished by early societies without the required infrastructure still remains not addressed and is largely unexplained. It has, also, been observed that some of these projects could not be duplicated with today's advances in technology and heavy equipment. Some megaliths in the hundreds of tons weight range, such as those found at Jupiter's Temple in Baalbeck in Lebanon, shown above, defy explanation as to how they were cut and moved.

Nevertheless, for the smaller structures; Stonehenge size and smaller, techniques have been demonstrated, not only in theory but practice. Mr. W.T. Wellington has managed to reproduce a Stonehenge like structure without hoisting equipment as shown in a video in his web site "The Forgotten Technology".

However, when it comes to the precision to which stones were cut and fitted to tolerances, achievable only with machinery, no explanation has been forthcoming. There are hundreds, if not thousands, of these magnificent joints in a single structure. The photographs below show two exquisite examples 2600 miles apart from each other, over the ocean and the Andes Mountains. The first one on is Ahu Vinapu at Easter Island, the other is a small example of the thousands of such stone fittings at Sacsayhuaman, Perú. The corner joints displayed at Sacsayhuaman are also found on the base of the Sphinx in Giza. How they were any of these accomplished is not known and defies explanation.

Above, AhuVinapu, EI, Chile   and   below Sacsayhuaman Cusco, Perú-
Wikimedia Commons

Another example of this exacting workmanship are the stone works at Puma Punku, Bolivia where cut stones are found in great quantity, having cutouts, circular holes and angles, stone work engineers have judged would be hard to achieve today. The first photo below shows one of these monoliths. The inlaid cuts are blind, do not go through. That makes it impossible to cut without a mill-like machine fitted with a diamond bit. This is not unique to these monuments. Precision stone cutting at its highest excellence is found in the Chephren and Luxor Temples in Egypt. At Luxor there are hundreds, if not thousands of glyphs carved on columns and obelisks that display astonishing design and precision; again not achievable with the tools of the time.

Wikimedia, Creative Commons photographs

In this chapter covered some of the basics regarding the characteristics, types, design and details archaeological sites may have, and that in some instances share with others, even when located remote from each other. In our analysis we used these elements to help define whether there was unity between sites. For this study the physical geometry of the site and its geographical location in relation to other sites were the underpinning characteristics we focused on. In some instances the overall design or design elements reaffirmed or emphasized the importance of the geometric arrangements found.

Notes

CHAPTER 2

A WORLD OF ALIGNMENTS, WHY?

 Setting aside for a moment the physical characteristics in design and the technological requirements needed to execute monumental projects, one needs to ask how building them was possible and to consider other possibilities as to who the authors of the archaeological record might have been. Also, need to keep in mind that some of the legacy may not be the product of our current civilization. Researchers have speculated an advanced civilization may have existed in an earlier period on earth; which they argue was extinguished by a cataclysmic event[18].

Another possibility, noted before, is the theory about alien technology from another planet, as has been claimed by many authors. Either way, in this study the best we could hope for, was to find evidence from our physical survey that would hopefully help us shed light on some of these possibilities, or perhaps favor one of them, if not provide a basis to conclusively reject some.

A technological aspect of civilization that may have been the result of, other than human ingenuity, which has received much attention from a mystical or energy fields perspective, are the alignments of monuments, burial sites, temples and others, even if located thousands of miles away from each other. The alignment of ancient structures has been claimed to exist, with other structures or with geographical landmarks, such as, Mount Ararat of Biblical importance.

Sir Alfred Watkins, in the 1920's conducted a cartographic study in southern England. He found ancient structures that aligned in a fairly straight line from near Minster in the east to Cornwall in the west; these results were published in the book; *Early British Trackways*. The alignments of ancient structures, megaliths, menhirs and circles, gave rise to his Lay Lines theory[8]. He claimed this was not a theory, since it was based on physical measurements. Others explained his measurements, giving rise to a theory that seeks to justify the positioning of these monuments in terms of energies. These theorists have moved from Watkins' physical measurements to energy fields' theories, which resulted in an orbital energy grid theory. A review of this transition and the evolving theory is available on Youtube[11].

There are proponents of other methods for finding alignments based on measurements. One of these methods, mentioned earlier, measures the distance from the earth's poles to define archaeological site alignments; Mr. Jim Alison, the proponent of method, describes a great circle that connects various archaeological sites from Easter Island to Cambodia, as well as, other mathematical and geometrical calculations that relate many sites according to their distance to the North Pole[12].

Author, Zachariah Sitching argued the alignments of the pyramids with Jupiter's temple at Baalbek and Mount Ararat and the temples of Jerusalem and Baalbek are the product of extraterrestrial visitors. The alignment part of his argument we considered would be easy to verify by the average individual with readily available mapping technology. Other theories such as the energy grids theory or other of Mr. Sitchin's theories which he claimed were also based on history decoded from the texts on the Enuma Elis tablets, would be nearly impossible to verify. One such claim -we leave to others to elucidate- is the alien's biological intervention in the creation of the human race, as he relates in the book The End of Days[2]. In this book, he describes how the extra terrestrial Annunaki conducted DNA experimentation to create man out of our hominid ancestors.

The geographic structure alignments Mr. Sitchin describes in his books, are the alignments of the Giza pyramids; Cheops and Chephren (30°N 31°W) with the temple of Jupiter at Baalbek (34°N 36°W) and Mount Ararat (39.7°N 44.3°E). Also, the alignment of the Sphinx on the 30th parallel (29.97°N, 31.04°W) with Basrah (30.53°N 47.8°W), and the alignment of the monolith platform under The Temple on the Mount at Jerusalem (31.77°N 35.21°E) , the Golan circle, (32.9°N 35.8°E) and Jupiter's temple at Baalbek (34°N 36°W). In the book he relates how these alignments, as well as others, were used as landing corridors for alien aircraft. The megalithic platforms under the temples, he explains, were the actual landing sites. One of these corridors, in particular, which we will discuss later, is the Mesopotamian corridor from the Persian Gulf through Basrah (Sumer) and Baghdad.
The idea that ancient structures are geographically aligned is an intriguing concept. If true, it would be reasonable to assume it is possible other measurable alignments should exist, in addition to those mentioned. If they were to be found, they would provide more credence to some of Mr. Sitchin's allegations.

However, the existence of alignments by themselves, were and will be critiqued on the basis that anything would line up by chance alone; particularly when the structures are found in areas where there is high Archaeological site density, such as in England, the Middle East or Mesoamerica.

These theories presented a good preamble to our research. To satisfy that need we sought to find other alignments, but first needed to test Sir Watkins' and Mr. Sitchin's claims. Before doing that, we felt a secondary standard beyond alignment was needed, such that it would add credibility to the results by eliminating or diminishing the probability of the alignment occurring by chance alone. We considered structure type, design features and more importantly the distance between the site and the line. In contrast to those who theorize about the energy grids, who, include everything in the alignment's path. Their rationale is that the sites were selected to absorb the energy from the fields they were placed in. Supposedly, the structures focus the energy thus becoming a 'watershed ' for those who are near these nodes. Earth grids have been extensively studied and the theories vary in degree of plausibility[11].

Before continuing, we need to define what we consider an alignment.

DEFINITION

We defined geographical site alignments, as a locus of structures which are near to or at a point on a straight arc line that connects them, within a specified maximum distance to the arc; we chose one arc-degree, approximately 69 miles at 90° from the starting point, at 45° half that distance and so on. *Every straight line drawn on the surface of the earth is a segment of a great circle-ellipse, 40,075km.* (24,901 miles) in circumference at the Equator.

To initiate the tests, we found the coordinates for the places mentioned by these two authors, then, traced the arc lines connecting them. It was confirmed that they do line up within a reasonable distance to the line within 10 miles thus proving both authors' assertions.

At the onset, we dismissed Mr. Sitchin's 'landing corridor' concept. The need for a landing corridor seemed to us a useless rationale. Alien technology advanced enough to get to earth would not need these alignments for landing purposes, we reasoned. The picture below shows the results of our test. Mr. Sitchin's corridors are highlighted in white. The line to the left starts at Giza and connects with Mount Ararat, marked with a dot marker at the top of the picture.   Approximately a third of the way up, the line passes over Jupiter's temple at Baalbek; this is the first alignment. The second line inside the triangle starts at Jerusalem crosses over the Golan circle ends up at Baalbek, this is the second alignment.

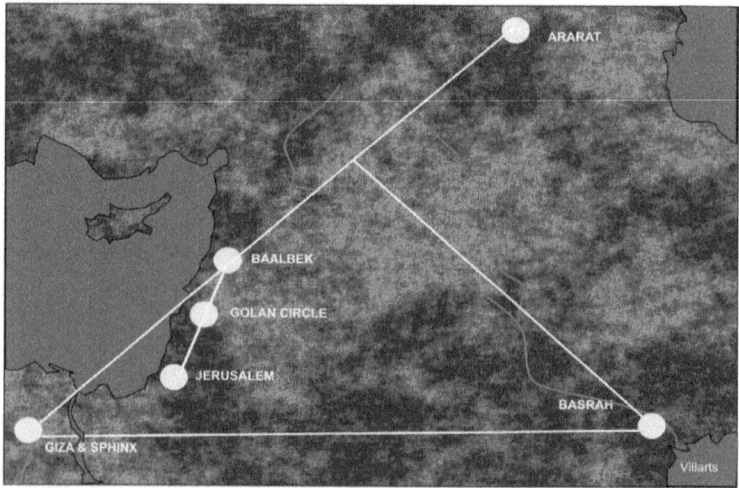

Mr. Sitchin also talks about an unidentified point south of Jerusalem, located at the point where the Baalbek -Jerusalem line would cross the horizontal line Giza-Basrah line. This point in the Sinai Peninsula, Mr. Sitchin tells us was the location of an alien military installation that was destroyed. The three lines connect three points, the minimum required for an alignment. The fourth line, in the north-west direction starting at the Persian Gulf through Basrah and connects with Baghdad, is the Mesopotamian corridor mentioned earlier.

During this exercise we found that the Temple on the mount, itself, is not aligned with the cardinal points, but, the dome aligns at 90°E with the Mount of Olives, where Jesus' Ascension is believed to have occurred, and with the Pater Noster Cathedral.

These facts heightened our interest, as they raised two issues: The cost of execution and the scant information which is available. We figured that to design the alignment and construction of structures of the magnitude of the pyramids at Giza, the Sphinx, the (base of) Temple on the Mount in Jerusalem, the Golan circle (not in the same category in complexity) and the (base of) Jupiter Temple at Baalbek, would drain the resources of the entire region if they had been built under a master plan. We estimated that today, for each structure, it would require about 10 to 25% of the combined gross domestic product of all three nations in the region; Egypt, Israel and Lebanon, whose combined 2010 GDP was about 135 billon USD. To provide some perspective, the original NYC World Trade Center took 7 years to complete at a cost, then, of about 1.5 billion dollars; in today's money it would cost from about 8.5 billion to 30 billion. (From the WTC web site). For the new WTC complex, the New York Daily News web site reports: "In 2008, the project was estimated to cost $11 billion to complete. The latest projection pegs the total bill to run close to $15 billion". Not including in the equation the cost savings due to current technology and the existing human resources and infrastructure.

A more contemporaneous example is the reconstruction of the Forbidden City in China by Emperor Zhu Di. Author Gavin Menzies in his book 1421 The Year China Discovered America[4] gives us a clear detailed picture of what it was like to carry out a project of great magnitude: It took seventeen years to complete, 1404-1421. The need for manpower and food to feed them he describes thusly;

> "A vast army of workers was also required to accomplish Zhu Di's vision and hundreds of thousands of Chinese labourers were force marched to the north; some 335 army divisions were re-deployed to guard them...but, feeding the first construction workers soon began to prove difficult. The growing season in the north was short; millet could be grown, but not rice, and corn and barley produced poor yields."
>
> "ON 2 FEBRUARY 1421, CHINA DWARFED EVERY NATION ON EARTH"

The second quote is the first sentence of Chapter One of Mr. Menzies book (His uppercase). Studies on the Population of China, 1368-1953[5], reports the population of China at 65 million people in 1420. In contrast, the whole Roman Empire in 25 BC had approximately 50 million. This provides a comparative scale as to what the available resources might have been back when the pyramids, Baalbek and the Temple on the Mount were built. The actual dates for when these monuments were built is not known.

The construction date for the pyramids is controversial. Some archaeologists believe they date thousands of years prior to the arrival of the Egyptian Pharos; which would make the infrastructure available to support this undertaking much smaller still.

With respect to information, besides Mr. Sitchin's corridor concept and Sir Watkins Lay Line theory, there are scant references found of similar claims for any other monuments or ancient sites that include actual measurements.

There are many references to studies regarding, earth grids, nodes, and other properties; these are given in the bibliography[13]. Armed with curiosity we set out to investigate whether other such measurable alignments indeed exist. Using readily available tools, such as; Google© earth and GeoHack (at wiki.toolserver), a survey was conducted of over one hundred prominent archaeological sites. We attempted to discern any patterns regarding their geographic location in relation to each other's geographical latitude and that of several naturally occurring prominent landmarks, as Mr. Sitchin had suggested. In our study we included other major peaks; included were Mounts: Ararat, Everest, Kilimanjaro, Chimborazo and Aconcagua. To initiate this study, the global coordinates for these major peaks and archaeological sites were tabulated.

Mr. Sitchin writes, regarding the reason for the DNA experimentation to create humans, was the need the aliens had for slaves to mine the ores they came to earth to exploit: gold and copper. Therefore, to round out the study, we included sites associated with the origins of man. Starting in Africa, where he described the mining first took place. We included the Bouri Peninsula in Ethiopia, where Homo sapiens, Ari Ramidus and Australopithecus were found[14], and Taung Child, in South Africa. To add global perspective, we also included in the list: Neanderthal man (Germany), Peking Man (China), Java Man (Indonesia), La Chapelle-aux-Saints, where "the Old Man" Neanderthalis was found (France), Monte Verde (Chile), the site of the oldest (14800 BP) human remains in the Americas were found. We also added other burial sites of interest such as: Lake Mungo in Australia, and the Island of Marawa in the UAE, where 6 thousand year old remains were recently found and included the prehistoric caves of; Altamira Spain, Chauvet and Lescaux in France.

During the course of our study, in October 2013, the oldest human remains were discovered in Dmanisi, Georgia; a complete skull, of a Pleistocene era human, named #5. It is 1.8 million years old - now re named: Homo erectus ergaster georgicus[10]. Its location was added to our list.

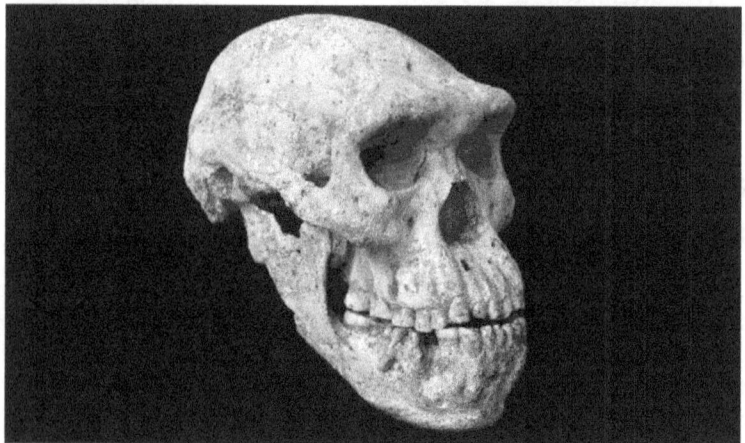

Skull #5,Dmanisi, Georgia. Image credit: Guram Bumbiashvili/ Georgian National Museum

In the first part of the study we compared 77 sites to find their geographic alignments by plotting each site location's latitude vs. longitude. This exercise yielded intriguing results. The plot is shown on the next page. In the plot it can be observed that the sites tend to group, in steps, most prominently around the 20th, 30th , 40th. and 50th. parallels in the Northern Hemisphere. In the Southern Hemisphere the number of sites are scant; with Monte Verde (41°30'S 73°12'W) in Chile, and Lake Mungo in Australia are the southernmost sites.

This hemisphere also includes some other important archaeological sites: Taung and Sterkfontein "The Cradle of Humankind" where an adult Australopithecine was found in 1936. Also, Tucume, Caral, Nasca, Easter Island, Machu Picchu, Tihuanaco and Puma Punku, the latter believed to be the oldest civilization in the Americas and the Ha'amonga Trilithon in the island of Tonga.

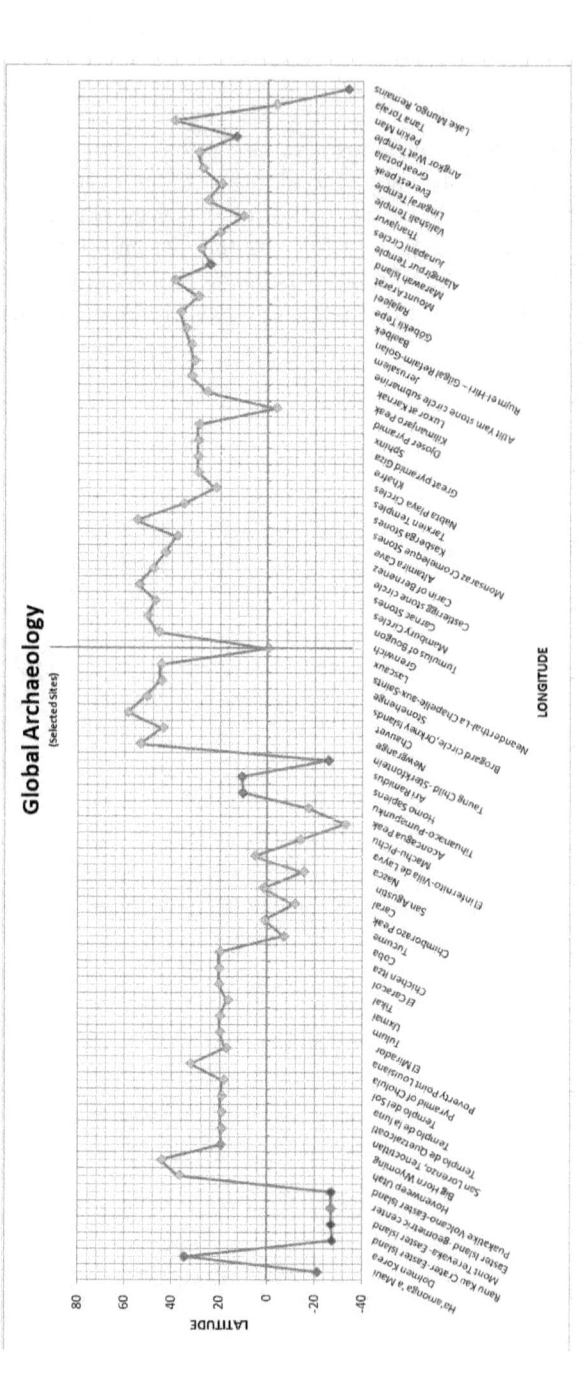

Global Archaeology
(Selected Sites)

The data did not yield prominent patterns between landmarks and archaeological sites. However, when some of the sites are connected by arc lines or great circles, in a manner similar to that employed to verify Mr. Sitchin's corridors, significant alignments were found. We identified at least twenty-four sets of sites comprising locations that lie on the same arc-line, according with the defined standard. Fourteen of these arc-lines connect three sites some are continents apart; that could be significant. More significant still, is that seven of these groupings connect four sites, one connects five sites, one seven sites and one seventeen sites!

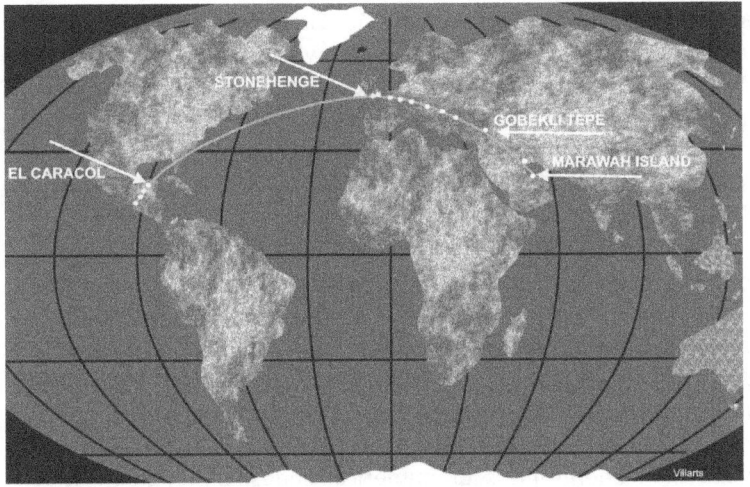

The Great Arc Alignment

The line connecting seventeen sites, we refer to as the Great Arc, is shown in the graphic above. The Great Arc 'starts' at the Citadel of Tontiná in the estate of Chiapas in Mexico. In the Yucatán peninsula aligns with the El Tígre Citadel, the Hochob Temple, the 'observatory' El Caracol and the Warrior's Temple in the latter two in Chichén Itza

Across the Atlantic in England, the arc passes near Avebury, Stonehenge, further away in Turkey it finds Göbekli Tepe. The latter, is over 7,500 miles away. Of these, three structures have similar designs consisting of concentric stone arrangements, confirmed or believed to be aligned with heavenly bodies; some archaeologists refer to them as 'observatories': El Caracol, Stonehenge and Göbekli Tepe. Göbekli Tepe has not been confirmed to have alignments with planets or stars.

The line, also, passes within 69 miles through: London, Brussels, Frankfort, Vienna, Budapest, Bucharest, Varna, Babylon, Baghdad, Basrah (Sumer on the 30th. parallel in Mesopotamia-the cradle of civilization) and ends at Abu Dhabi (near Marawah Island). This great arc runs in the direction of Mr. Sitchin's Mesopotamian landing corridor previously mentioned, which he described as starting at the Persian Gulf. All these cities contain or are nearby archaeological sites of significance. Varna in Bulgaria, for example, is the site for the world's oldest gold mine, estimated to date back to from 4600 - 4100BC. Some of the others are, the Neolithic settlements: Praunheim, Höchst, Niedereschbach, near Frankfort, and the Gumelita civilization in Bucharest, dating back to 4500BC; also, in the middle of London recently an ancient Roman structure was discovered, dated 40AD. A list of sites connected with arc-lines appears in the table below. These two data sets give some credibility to the argument of existing theories that claim a global arrangement or a global plan; be it grids or numerical codes[11,28]. These alignments, we estimate could have a more practical purpose than energy grids or numerical codes; although these could reinforce some aspect not yet guessed at.

Alignments within the same country could be dismissed as coincidental, as were Sir Watkins' Lay line measurements. Same or similar cultures, with locations within a few hundred miles of each other, could be argued, 'easily' account for the similarities in these structures and their locations casually falling in line, particularly if

Great Circles	# Sites	Location
Caracol-Stonehenge-Göbekli-Babylon-Basrah-Marawa*	17	Mexico - England-Turkey-Iraq-UAE
Mirador-Calakmul-El Castillo-Stonehenge-Sipiennes-Varna-Göbekli	7	Guatemala-Mexico-UK-France-Bulgaria-Turkey
Tikal-Kohunlich-Chachoben-Mambury-Basrah	5	Guatemala - Mexico - England - Iraq
Easter Island-Tucume-Gobekli-Junapani	4	Chile - Peru - Turkey- India
Carnac-Tumulus-Lascaux-Tarxien	4	France - France - France - Malta
Tikal-Tucume-Caral-Nasca	4	Guatemala - Peru - Peru - Peru
Teotihuacan-Cholula-Monte Alban-Puma Punku	4	Mexico - Mexico - Mexico - Bolivia
Teotihuacan-PovPT-Callanish-Göbekli	4	Mexico - US - Scottland - Turkey
Ha'amonga 'a, Tonga-Potala-Bagdad-Jerusalem	4	Tonga - Tibet - Iraq - Israel
Easter Island-Machupicchu-Giza	3	Chile - Peru - Egypt
Godmanchester-Stonehenge-Mambury	3	England - England - England
Castlerigg-Stonehenge-Tumulus of Bougon	3	England - England - France
Stonehenge-Carnac-Altamira	3	England - France - Spain
Tikal-NorthPole-Great Potala	3	Guatemala - North Pole - Tibet
Teotihuacan-Machupicchu-Puma Punku	3	Mexico - Peru - Bolivia
Nasca lines-Stonehenge-Great Potala	3	Peru - England - Tibet
Machu picchu-Chimborazo-Tikal	3	Peru - Ecuador - Guatemala
Machupicchu-Giza-Potala	3	Peru - Egypt- Tibet
Nazca line-Machupichu-Baalbek	3	Peru - Peru - Israel
Brodgar-Carin-Altamira	3	Scottland - France - Spain
Kasberga, åles stenar-Tumulus-Altamira	3	Sweeden - France - Spain
Ha'amonga 'a , Tonga-Easter Island-Puma Punku	3	Tonga - Chile - Bolivia
Ha'amonga 'a, Tonga-Potala-Giza	3	Tonga - Tibet - Egypt

*Also: Hochob, El tigre, Tontiná, Warrior's Temple, London, Brussels, Frakfort, Vienna, Budapest, Bucharest, Varna, Abu Dhabi they have an astronomical reason for being where they are located. In the table below, the pyramid cluster locations in the Yucatán Peninsula and those in Tenochtitlan, Teotihuacan and Tula, which are also, within a hundred miles of each other, are examples. However, in the first entry in the table, as mentioned, the distance between El Caracol in the Yucatán peninsula, Stonehenge in England, and Göbekli Tepe in Turkey is over 7,500 miles and to Abu Dhabi 8,700. Half a great circle at the equator is approximately 12,500 miles as far a distance as one could go on earth.

At first glance it seems hard to put together a rationale that would explain the Great circle alignment beyond just chance. It could be claimed that without a technology capable of global reach in ancient times, at least as sophisticated as Google© earth, the alignment could not have been planned from earth. However, with a tool with this capability; a guidance system on a geo stationary ship (adopting Mr. Sitchin's account) and GPS, the locations for the ancient settlements where today's cities and the structures are found, could have been planned. See graphic above. For this purpose, the positioning of such satellite ship would have had to be known to communicate with it. Here is where the Analemma for the ship (the Infinity symbol) discussed earlier would have been useful, and the mystical/religious connection would also make sense.

Nevertheless, for the first arc-line on the table, it can be speculated the reason for these locations to align in this manner, might be that the original settlements were selected serendipitously. Ancient peoples, in different places may have founded these locations to view the same astral phenomena, from different points, even at different times.

The angle for the Great Arc connecting the 17 locations is about 40.5° which is roughly the path Venus appears in the sky. When the planet passes in front of the sun - its transit, -see NASA's photo below- the phenomena could have been interpreted as a signal from a deity.

NASA's_SDO_Satellite_Captures_First_Image_of_2012_Venus_Transit

Venus' transit path is shown on the graph below as was seen from earth on June 5-6, 2012. The Great Arc is shown as semicircle line superimposed onto NASA's graphic of the Transition (The Kennedy Space Center, coincidentally, happens to be located in the Great Arc's path). The Mayan civilization called Venus: Noh Ek "The great Star"; the Caracol "observatory" has been shown to be aligned with Venus[32]   Sir Norman Lockyer, an early Stonehenge astroarchaeologist did not find alignments for any of the so called, naked-eye planets; Venus being one of them[15].

The second alignment line on the table which connects seven archaeological sites runs almost parallel to the previous line. Unlike the previous line, this may have been developed serendipitously tracking Venus; of the sites on this line only two are observatories: Stonehenge and Göbekli Tepe, three are pyramids and the last two are ancient mining sites. No references were found that would substantiate whether these pyramids have specific alignments with Venus, although, it is likely. The balance of the site alignments on the table has not been explained.

FIGURE 1

Global Visibility of the Transit of Venus of 2012 June 05/06

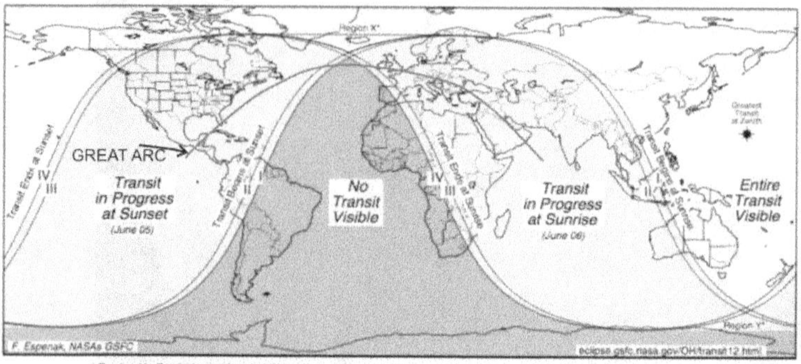

* Region X - Beginning and end of Transit are visible, but the Sun sets for a short period around maximum transit.
* Region Y - Beginning and end of Transit are NOT visible, but the Sun rises for a short period around maximum transit.

In a follow up study, we pursue what appears to be another enigma: Why so many locations align in this manner. Absent some rationale that explains it we cannot entirely, dismiss at this point alien visitation and, perhaps, colonization. The judicious layout of structures, citadels and ancient settlements would indicate a long term plan. Short term visits would not warrant such vast enterprise, if that in fact that was the case.

This part of the study has broadened the field of our intended inquiry and in an unsuspected way given more credibility to the arguments that remove our current civilization as the authors of these monumental works. The most likely possibilities remaining are; earth visitation by extra terrestrials who may have assisted in the development, or the earlier existence of advanced civilization(s) on earth. In the second part we expand the scope of the study of alignments looking for an explanation of the findings in a global context.

Notes

CHAPTER 3

THE CANDELABRA

In the previous chapter we talked about
archaeological sites and locations where human
history may have started, some which were found
are aligned to great circles drawn around the earth.
We confirmed the claims two authors had made
regarding these alignments and expanded the
number of alignments they had found. Reviewing;
we found twenty four sets of aligned locations:
there are fourteen alignments, each with three
sites, seven with four, one with five and one with
seventeen sites. Assessing these results, we
questioned the need for logistics, man-power and
the economic enterprise that would have been
required to carry out those projects and the
magnitude of the enterprise that would have been
required to complete them.

Although we presented an explanation for some of the alignments, the majority of them went unexplained. Originally we started out to prove some archaeological sites are aligned as had been claimed; we did not expect to greatly improve on the number of alignments found to twenty three additional alignments or expand the number of sites in the Basrah-Baghdad corridor to 17 sites reaching as far as the Yucatan peninsula in Mesoamerica and almost ending at the Pacific Ocean!

The proponents of grid theories believe thousands of alignments exist that follow the grid. This discovery tilted our balance in favor of the possibility of earlier technically advanced civilizations having lived on earth, capable of the design and construction of these alignments. To reach that tentative conclusion, we pointed out the need for a satellite based global positioning system for those sites which do not track a planet's trajectory, and how the mathematics represented by the spirals and lemniscates found everywhere, could support the claim the ancients had knowledge of space flight and communications.

For that part of the research, as we traced the lines between sites, we kept our pre determined criteria to judge whether a site was in line or not. This worked fairly well for most sites. We placed the markers as close to the center of the location or structure; for a circle, we picked the center and for a pyramid its apex. That strategy was simple and straight forward and it worked, until we got to Nasca.

At first we thought of using Old Nasca town as the site for the marker, but that seemed skewed to the south of the actual plateau where most of the glyphs and lines are found. Although the plateau itself would fit within the constraints of our one degree parameter, somehow it seemed arbitrary to jus pick a central point.

As we surveyed the plateau's surface we became aware of two lines that bisect the area into four fairly equal quadrants. These lines, each are about 6 miles long, and are the longest lines in the plateau. We selected their intersection point to set the marker for the site.

As we became familiarized with this unusual place, learning its history and poring through archaeologists' studies, conclusions and explanations; its uniqueness in design or lack of it struck us as being pointedly odd in comparison to the other one hundred plus sites we had surveyed. In the end we found the site belongs to what we consider a unique group comprising other similarly mysterious places and their objects: Easter Island's Moais, Verneukpan's spirals, in South Africa, El Infiernito in Colombia with its phallic megalithic arrangement, Area 51's geometric glyphs, the circles of Junapani and Sine and the Sajama lines and kivas and Kirbeth's 'kites' and Lemniscates.

Nasca stands out in its apparent scattered disarray of lines and gigantic figures that look like an infant's painting; however, the figures themselves are exquisite in design and drawing technique. The Colibri -Hummingbird is shown below. The figures are drawn with a single line that does not cross itself and their sizes run in the hundreds of feet; except for one: The Astronaut.

In the picture below, on the left is the Nasca Astronaut, on the right is a glyph found at Butler Wash, UT, US.

The other sites in the group of mysterious places follow design themes with differing motifs; Easter Island's Moais, placed on pedestals called Ahus, are found in coves on the island's shores. Verneukpan's spirals most of them are arranged in rows inside an inverted 'V' and a few dispersed groups outside of the V and Area 51's triangle, quarter circle, six point star and fork; are standard geometric figures, as are the two Bull's Eye. At first we were reluctant to include the glyphs in Area 51 in the study we had reservations as to their archaeological value. We will review this site last. The sites of Sajama and Kirbrth Al Umbashi will form part of a separate study, due to the need to take a closer look at the similarities they appear to share.

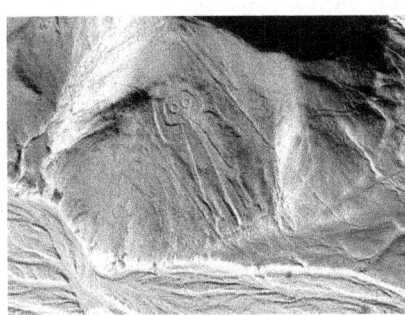

Wikimedia Commons

Of the hundreds of known archaeological locations, these are few that we consider belong in a unique group stand out like the proverbial 'sore thumb'. Their individual designs are like no others on earth, although some of the other sites may contain similar tie-in elements, such as: Spirals, Lemniscates, triangles, circles, stars and trapezoidal figures, whether they are found as decorations by themselves or as parts of large structures.

At first we thought, Nasca was a capricious array of geo glyphs; that was our point of view until we begun looking at the Nasca plain in more detail and from a different perspective. Perusing a popular archaeology survey book; Mystic Places[6] we came across a full page photograph of the famed glyph 'The Candelabra'.

The Candelabra's design, size and location capture the imagination: why is this glyph is so far apart from the others? Its design is so different in technique -no single line like the others on the Nasca plain, or the Astronaut glyph; it warranted further study. This strange figure has been described in many fantastic ways. A fanciful one we quote from Science Frontiers web site: "F. Joseph, thinks it looks like a Jimson weed! Furthermore, he states that there is a miniature version of the Candelabra drawn on a rock in California's Cleveland National Forest. Joseph associates the two candelabras in this way: The ancient inhabitants of Perú voyaged to California to collect Jimson weed and other hallucinatory drugs. When they sailed back to Perú with their cargo, they used the Pisco geoglyph as a navigational aid!

THE CANDELABRA

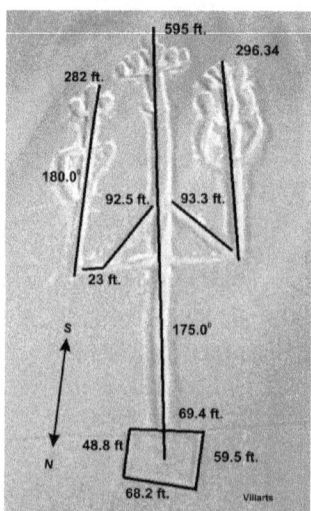

We begin our study by defining the Candelabra's geometry and positioning with respect to its geographical location (13.794°S 76.308°W). As in the previous study we began searching for alignments it could possibly have with other locations. The results were incomprehensible at first, until they yielded what appears to be a grand design. We begin to unravel its meaning by explaining the Candelabra's relationship with Nasca. In one respect we agree with Mr. Frank.

The Nasca plain's location is not obvious to travelers, even from short distances is not esily seen, there are no prominent landmarks; the glyphs lay on a virtually horizontal plateau. It appears it was necessary to mark its location for newcomers to find it approaching from a distance. Approaching the Nasca plateau from the ocean from the NW, the Candelabra, is a welcoming sign. It is carved as a 'guide post' at the shore in Paracas pointing to it.

It is reported the Candelabra can be seen from 12 miles out to sea. It is an isolated geo glyph about 170 meters (595 feet) long carved on the north western face of a low mountain at the tip of a peninsula, facing the ocean like a billboard - in contrast to the other geo glyphs at Nasca, which are located some one hundred miles away and lie almost flat and can be easily missed (discovered in 1927). It does not depict an animal or any other known object. Its shape appears to be a sophisticated pointer; not a Candelabra, as Von Däniken had described it in "Chariot of the Gods". Von Däniken suggested it looked like a 'pointer', but never found out what it is pointed at. We measured its bearing and found the Candelabra's main prong points south at ~175° to the tip end of South America; Tierra del Fuego. Before reaching the end, just east of Puerto Montt in Monte Verde creek in Chile finds the location of the oldest (14,800BP) pre-Clovis human remains found in the Americas (41°30'S 73°12'W). In the same direction, just over one hundred miles south-east of Paracas is the location of the Nasca plain; the site of the famous lines and anthropomorphic glyphs.

The Candelabra has three arms. Two short arms on either side of the main shaft -see drawing below- forming a trident fork like figure: the 'pointer'. The left arm points to the geographical South at 180°, the other two are parallel pointing at 175°. This small 5° difference shows purpose in its design. Atop each arm; right, center and left, there is a 'crown'. It is obvious the prongs were designed to be dissimilar and asymmetrical, as well. We believe this asymmetry may encode other properties beyond setting directions to locations of interest.

The design does not appear to be capricious. The crown on each arm has a different number of lobe-like endings. The left arm has two lobes on either side of its stem. Counting the number of lobes on each stem, we totaled 17, broken down 2-8-7. We speculate this graphic represents the electron arrangement of chlorine: 2s 8p 7d, and the lone small lobe below the crown on the main stem, probably represents Hydrogen's 1s electron. The chloride ion is vital to human existence. That's one possibility. The other, is perhaps, a better interpretation: Oxygen's atomic number is 8 and Nitrogen's is 7; these are the principal gases in the atmosphere. Molecular Hydrogen is H_2, plus Oxygen (a single lobe below the oxygen arm) together makes up water; as vapor or present on the planet. The Candelabra records the most important elements in our atmosphere and planet.

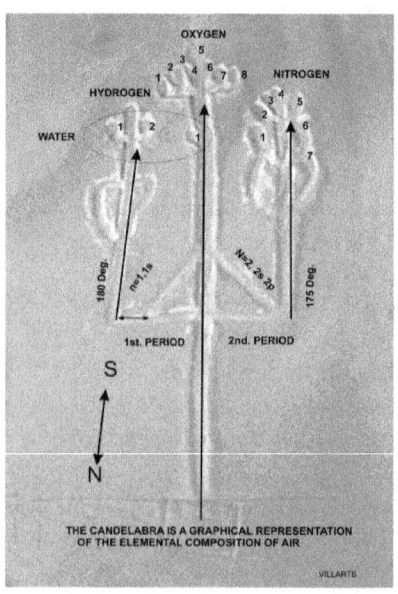

THE CANDELABRA IS A GRAPHICAL REPRESENTATION OF THE ELEMENTAL COMPOSITION OF AIR

The periodicity of the elements was, also, known to the designers: The glyph shows the Oxygen and Nitrogen stems together, closely connected by a cross member at the base, while the Hydrogen stem is set '23m apart' from the cross member; emphasizing the difference in their principal quantum numbers: Hydrogen n=1 1s, Oxygen n=2 2s 6p, Nitrogen n=2 2s 5p; the latter have two principal quantum numbers n=1, n=2.

This is emphasized by the difference in the stems' azimuths: Hydrogen's 180°, while the other two prongs representing Oxygen and Nitrogen have the same bearing: 175°

The Candelabra is a graphical representation of the elemental composition of air; part of the archaeological record of an advanced civilization. The above information would be a good thing to know if you came from outer space. Although, one can argue, astronauts would have had the capability to know this a priory, before entering the atmosphere, via absorption spectra or any of other instrumental techniques. It is surmised that knowing this information, in itself, could not have been their concern.

If an earlier advanced civilization existed on earth, later in the argument it will become apparent they had an overwhelming need or desire to, literally, cast the record in stone.

After it was discovered the symbol has information encoded in its design; the atmospheric composition and the alignment of its geographical position and azimuth; we hypothesized, every line in it would point in a specific direction, as well. The support for the stem on the west side of the figure is found to have an angle of 127.50°, thus it points in a SE direction. A line drawn from it at this angle over the mainland crosses over the remains of various unidentified settlements, until it reaches the lost city of Huayuri and the geo glyphs of Llipata before reaching the Nasca plateau at the top and continuing South East where it reaches the ancient settlement; Tambo de Zapahuira.

Further south it crosses the Salt flats of Surire, Coipasa and Uyuni the largest salt desert in the world; all important sources for salt-NaCl -Sodium and Chlorine and also Nitrate. (Nitrate is needed for space flight since it is used in rocket fuels.)

The support for the stem on the east side of the figure has a bearing of about 38.56° NE, a line in that direction, in Guyana crosses over one of the richest gold regions in the world; Cuyuni-Mazaruni (6.46°N 60.2°W).

Continuing across the Atlantic, in Europe, the line crosses over the Paleolithic Flint Mines of Sipiennes and near Erkrath, Germany where Neanderthal man was found. In Tibet it finds the Dsong Citadel in Gyantse (28.94°N 89.6°W). In Thailand the line passes over the temples of Chiang Dao, Prasat Huay Khaen,

Phanomrung, also, the preserved remains of Prasat Thamo and in the border with Cambodia Prasat Ta Muan Thom (14.35°N 103.27°E).

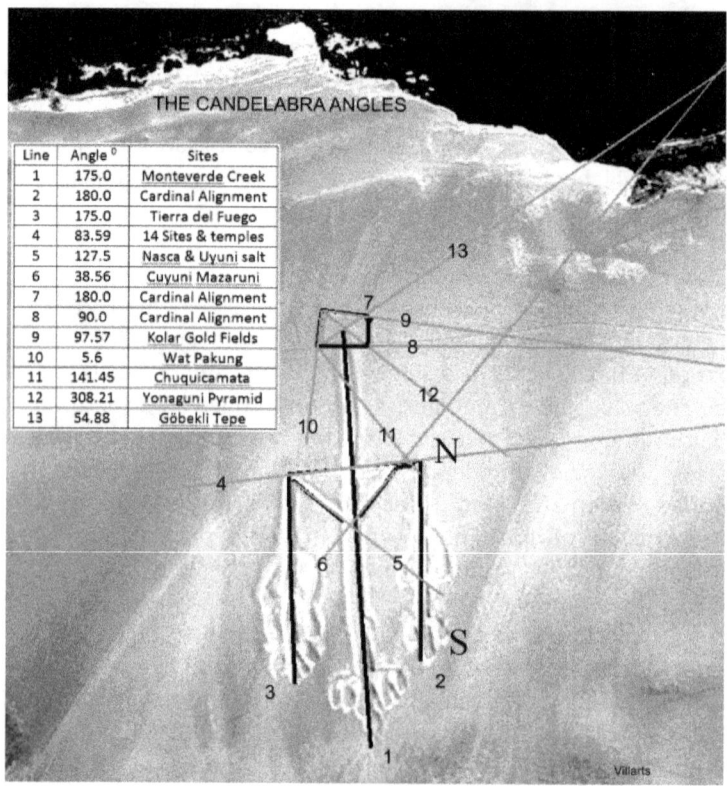

THE CANDELABRA ANGLES

Line	Angle°	Sites
1	175.0	Monteverde Creek
2	180.0	Cardinal Alignment
3	175.0	Tierra del Fuego
4	83.59	14 Sites & temples
5	127.5	Nasca & Uyuni salt
6	38.56	Cuyuni Mazaruni
7	180.0	Cardinal Alignment
8	90.0	Cardinal Alignment
9	97.57	Kolar Gold Fields
10	5.6	Wat Pakung
11	141.45	Chuquicamata
12	308.21	Yonaguni Pyramid
13	54.88	Göbekli Tepe

Graphic by the Author over a Google© Map

A line starting from the south west corner of the quadrangle base, in a southeast direction to the Candelabra's support aligns at 141.45°SE with the base of the stem on the east side.

Following the line in this direction we find the world largest copper mine Chuquicamata, in Chile. The quadrangle's top line runs east at about 97.57° E. Following this line to Africa it crosses the province of Katanga north of the African copper belt. Further east in India passes thirty miles south of the Kolar Gold fields. Gold sourced from this region was found in Mohenjo Daro in the Indus valley; dating its mining as early as 1000BC[23] The North-East diagonal line across the base with a heading of 54.88°, points to Göbekli Tepe in Turkey, currently dated at about 11-12k years BC.

> Recalling Mr. Sitching's argument; the reason extraterrestrials selected to visit earth was its resources, such as gold and copper; it appears that other resources would also had been of interest. We also found: flint, iron, salt, bauxite and uranium. The building materials, perhaps were needed as raw materials to be used locally, or to be used for future survival, except for gold, or maybe. It has been claimed the aliens wanted a specific type of gold; its mono atomic form, found in the Persian Gulf and claimed to have unique physical and biological powers such as imparting humans with longevity, as has been presented by reporter Jim Marrs[16] and others.
>
> Flint was an important raw material used since Neolithic times in diverse applications; for buildings and as fire starter. The first Great Arc-line described in Chapter 2, crosses near Brussels. Just south of Brussels, at Sipiennes is a Paleolithic flint mine. The line also crosses below Grimes Graves in the UK, another flint source.

The Candelabra's trident's base line has an E-W trajectory of 83.59° - 263.59°. A great circle traced at this angle, in its eastward direction crosses over the Tambo Colorado region, where the Inca Tambo Citadel and the Band of Holes are found (13.7°1N 75.82°W). Further east it encounters the Moray Inca and the Ollantaytambo Citadel. The Moray Inca, pictured on page 17, is a site with four below grade amphitheaters, two of these are joined forming a large lemniscates shape. Some researchers speculate the terraces were used for agricultural purposes.

Ollantaytambo is an Inca fortress city used by the Inca emperor 'Manco' at the time of the Spanish conquest.

The direction of one of the main walls atop the citadel runs south at 138.19⁰, following this direction, near Lake Titicaca the line crosses over the Aramu Muru Hayu Marka gate, a ceremonial site since ancient times.

Continuing east the ellipse crosses over India's west coast where it finds the temples of: Demashan, Tamdi Surla, Sri Amruteshwar, Hanuman, Malyavanta and Hampi.

In Thailand finds the temples of Prasat Thap Siam and Wat Nong Bo. Approaching Thailand from the east (263.59°) on the same circle are the temples of Tháp Nhạn and Prasat Koh Ker. Most of these temples are aligned with the cardinal points with their longest side of the structures in an east west direction. Great circles drawn over each temple at 270⁰, in Peru cut across over every major archaeological site in the Inca region. Of particular importance is the alignment of the Anantha Shayana 15.277⁰N 76.407⁰W Temple whose great circle goes across Machu Picchu and nine miles north of the Candelabra. In Thailand the Candelabra's trident base line passes fourteen miles south of the Ayutthaya temple complex. The main temple has an azimuth of 264.23⁰. A circle at this angle reaches the Nasca plain across the Pelican and the Astronaut glyphs.

The Candelabra's Circle of illumination (defined on page 71) at the summer solstice in the southern hemisphere, has an azimuth of 335.85⁰. A circle drawn at this angle follows the Peruvian coast line up to the border with Ecuador. Aligned on this line north of the Candelabra, in Perú are nineteen major archaeological sites: Incahuasi, Pocoto, the Huacas of Pucllana, Garagay and Paraiso and Cerro Culebras. One hundred miles north on the line is the Inca Fortaleza de Paramonga, eighty five miles further are the archaeological sites of Sechin, Pampa de las Llamas and Casma. Another hundred miles north are the Huacas Luna Y Sol and Arco iris and the Citdel of Chan Chan. Further north are the Canoncillo Ruins, Siete techos, Huaca Rajada and the Pyramids of Túcume. Continuing north across the Pacific Ocean, in Honduras and Guatemala are the Mayan sites of Copan, Los Sapos and Quirigua.

In the Yucatan peninsula the El Tigre citadel is also found. Across the Gulf of Mexico into the US, in the State of Wyoming the circle cuts over the Medicine Wheel -shown on page 26, in Montana finds the Pictograph cave and crossing over into Canada the Writing on the Stone archaeological site is found.

 The cartographic connection between the Candelabra's design and the remoteness and importance of the targets embedded in the direction of its lines, strongly suggest a grander design than just a primitive society's dabbling in art. The proposed hypothesis was satisfied, the Candelabra's design encodes directions like a navigational map. The target locations just described are but a few examples of many others that were found elsewhere, which will be discussed later. For the moment, among the many important locations the Candelabra points to, we focus on the Nasca plain, famous for its lines and glyphs. The Candelabra is one 'guide post' leading to Nasca. Approaching the coast of Chile from the SW there is another guide post: Easter Island. Its connection with Nasca is similar to the Candelabra's and will be discussed in detail later.

Notes

CHAPTER 4

THE NASCA REGION

The Nasca lines, have been named The Glyphs of the Gods by many; they have perplexed the world of archaeology since their discovery by Perúvian archaeologist Toribio Mejia Xesspe, in 1927 In the book Mystic Places the author ponders: "We may never know why or for whom the ground dwellers of the Nasca plateau or of Britain's hills created these earth drawings" [6] This quote reflects the old and the current view, and most generally accepted belief regarding the lines; even though many scientific explanations have been offered, most of them conclude the lines are a product of the Nasca peoples.

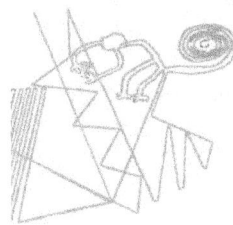

Not all agree; there are a few authors who have attributed the lines to ancient astronauts, or an ancient pre historic civilization as previously discussed.

From a technical perspective, studies have been conducted by scientists of virtually all disciplines yielding a plethora of theories with various degrees of vagueness as to the design and purpose for the lines: They point to water sources, they are ceremonial, and as a method to track stars and planets, as proposed by Maria Reiche a German mathematician and archaeologist who produced, perhaps, one of the most detailed studies and whose maps became the stepping stones for Nasca lines research. She concluded the lines are linked to astronomical events they show the rise of stars and planets and or track water sources (1938-1941). Her conclusions are still being debated. She is considered a celebrity in the city of Nasca where her former home is now a museum.

Considering the results of the first part of the study, in light of the Candelabra discovery -a glyph which encodes in its design directions to remote places- indicated that to study the Nasca lines, the method for their analysis should be the same as that we used for the Candelabra: *Follow the lines.*

THE NASCA PLAIN

Armed with and encouraged by the results obtained with the Candelabra's analysis, we applied the *'Follow-the-Lines'* methodology to the study of the Nasca lines. The rationale behind the method made sense. A line serves only a limited number of purposes: it sets a boundary, it connects two points, it measures the distance between them or it *points in a direction.*

The preponderant long lines on the Nasca plain, we had found earlier, were selected to initiate the study. Their azimuths were measured and ellipses (The earth is not spherical therefore circles drawn around it are slightly elliptical) were traced around the globe in those directions. One of the lines runs at about 324.86° NW and the other line 68.67° NE.

These two angle values are similar to the earth's axis tilt and the angle of the ecliptic at the summer solstice in the southern hemisphere. The earth's celestial axis is tilted 23.44° off the perpendicular to the orbital plane -the ecliptic.

At the winter solstice the axis' tilt angle is negative 23.44° or 336.56°, the explementary angle[47] with respect to the ecliptic plane, and the corresponding ecliptic Equator has an azimuth of 66.56°(azimuth 0°N). At the summer solstice in the northern hemisphere, the sun rays fall perpendicular to the axis tilted at 336.56°. The axis tilt varies, "plus or minus 1.3 degrees from its average value of 23.3 degrees. This number is not absolutely stable - it depends on the combined positions of all the planets through time"[46]. The two white lines shown -see graphic below- cross at the center of the Nasca plateau dividing it into quadrants, the gray lines are the earth's celestial axis and Equator.

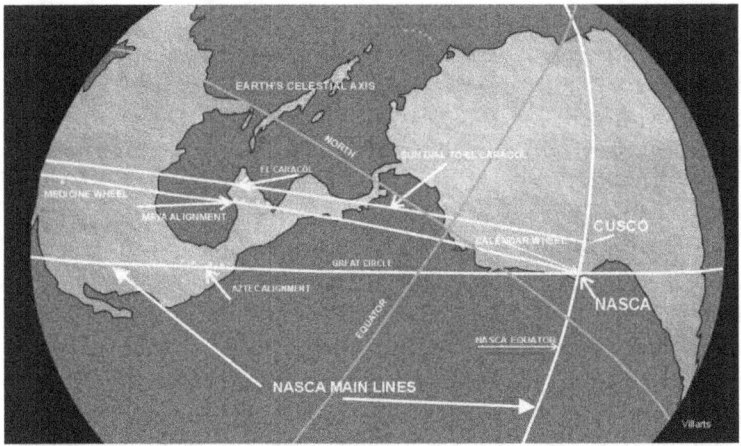

Following the line in the 324.86° NW direction, it connects Nasca with archaeological sites in México: Monte Alban, Cholula, Tenochtitlan, Teotihuacan and Tula; we named it The Aztec Alignment -shown in the graphic as a horizontal white line. In the US the line passes near four important sites where petroglyphs are found: Deer Valley in Arizona, the Atlatl Rock, Grimes Point in Nevada and Lake Winnemucca where the petro glyphs are estimated to be 10,000 year old.

Continuing the line onto China it finds the Temple Jinshan in Zhenjiang, further south in Silun, Guagdong province it passes by the Astronaut Rock[9]. Carved on this rock is a glyph that resembles the gold artifact shown on page 13 and one in the Chaco and Chelly Canyon glyphs in the US.

On the same line, just north of Nasca, there are three other archaeological sites of interest: The Band of Holes, discussed in Chapter 1, El Paraiso Ruins and the Huacho Salt Flats.

Following the circle south of Nasca in a 144.86°SE direction, it encounters the Whale glyph, Old Nasca, several geo glyphs at El Camino, the salt flats at Atacama, and in Cambodia finds the world's largest religious temple: Angkor Wat. A total of sixteen sites in a combined distance of 16,834 miles, three quarters of the way around the earth! The region, north Cambodia- south Thailand is the antipode to Nasca. All the Nasca and the Candelabra's lines cross there, each completing a great circle when extended around the globe in either direction.

As noted before, these two lines which cut the Nasca plain into four quadrants, when extended to the antipode region in Thailand; divide the world into four wedges. The two lines are not perpendicular to each other, so it appears they were meant to outline specific regions on earth. This skewed arrangement frames each region comprising a geographic commonality. The NE wedge encompasses nearly all the northern hemisphere's land mass and most archeological sites. The NW wedge comprises the Pacific Ocean. The SW comprises southern South America, Antarctica, Australia and most of Polynesia. The SE wedge comprises mostly jungles, the Amazon, southern Africa and southern India.

Another line of interest on the plain is a ray of one of the many that converge at a point in the Nasca plateau located at (14.697°S 75.135°W). Some archaeologists refer to these converging lines as 'stars'. This line is shown on the map (lower parallel line), on page 71' it has a bearing of about 336.84°NW. This is approximately the same value of the angle of declination of the earth's axis (+0.27°), and is nearly perpendicular to the 'ecliptic Equator' -short line between the paired lines, in the graphic above which connects Nasca with Cusco.

Here we pause the discussion to introduce the concept of the *Circle of illumination*. Understanding its mechanics provides the background to understanding the significance of the angle measurements that are found in archaeological structures.

THE CIRCLE OF ILLUMINATION

The *Circle of Illumination* is a line traced by the sunlight around the earth as it spins, this line separates day from night. Since the earth's axis is tilted the line of light does not go through the earth's poles; it goes around the earth as an ellipse at an angle of 23.44°. This is the same angle between the celestial equator and the ecliptic. As the earth orbits around the sun, the sun's location with respect to earth shifts position between ±23.44°.

The Nasca line on the plain at an angle (azimuth) of 336.84° is close in value to the declination angle of the earth's axis and

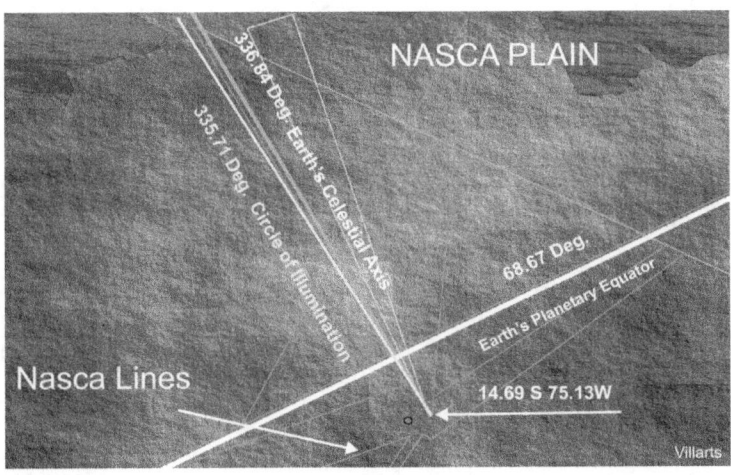

and to the azimuth for the *Circle of Illumination* at that location. At Nasca's latitude its value is 335.71° This line is perpendicular to the sun angle at the summer solstice; both lines are shown in the graphic above, as they appear on the Nasca plain.

Structures or places lined up along ellipses at the angle of the *Circle of Illumination* receive sunlight at the same time, i.e. the sun rises and sets at the same solar time -assuming there are no obstructions.

The Arctic and Antarctic Circles are imaginary circles formed by the set of ellipses drawn by the *Circle of Illumination* as the earth spins. Likewise, the *'ecliptic Equators'* -ellipses drawn perpendicular to the *Circles of Illumination* at a point equidistant from the poles generate the imaginary parallel lines known as the Tropics of Cancer and Capricorn circles. These circles represent the maximum solar altitude at the summer Solstices in both hemispheres respectively. When the sun is on the same plain of the Equator at 0°, the Vernal and Autumnal equinoxes shift from one hemisphere to the other-see graphic below. We'll revisit this subject later on in Chapter 6.

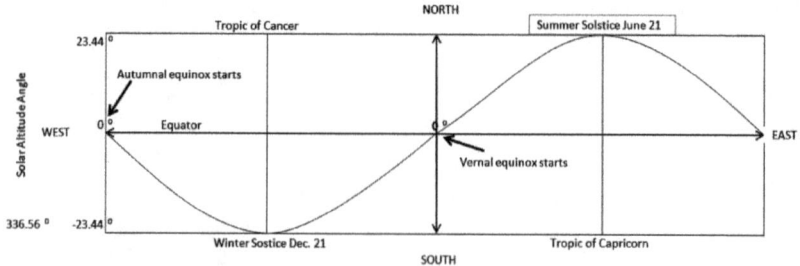

Following the Nasca line 336.84°NW northward, it crosses significant archaeological sites: On its way north, it finds a site made up of concentric stone circles, high up in the Andes at 14,647 ft. above sea level (12.416°S 76.15°W) we dubbed them the 'Andes Circles'. This structure has not been cataloged under any name. It was discovered, or for the first time described by Alberto Temple del Valle and Hugo Huertas, who explored the site in June 2009. They described it as follow:

> "...at the site they are marked as megalithic structures (Sambolin 2007). We reviewed the historical record and did not find references or studies or bibliography of any kind regarding the site, also, it is not classified as a national monument. This structure is not completely circular both circles are rather oval in shape. The exterior circle, measured from east to west is 148 meters and the internal diameter 84 meters; north to south it is 136 and 69 meters respectively."
> Translated by the author from their comments in Panoramio©.

Also, found along this line are, the Japani ruins, and the Yanacocha Gold mine (6.98°S 78.5°) in Perú, and in the Yucatán peninsula, México; it crosses less than ten miles from the
Pyramids of: Hormigueros, Becan and Xpuhil (18.54°N 89.43°), Hochob (19.42°N 89.77°W), Kanki K'nish (19.998°N 90.111°W) and Xcallumkin (20.17° 09.01°W). In Honduras it passes three miles east of the archaeological site of Cerro Palenque. We named this line The Mayan Great Arc; also shown in the globe map above; the lower of the 'parallel' lines.
Again, the sun raises and sets at all these sites at the same time due to their alignment; they mark the *line of sunlight*, like pins on a map marking a route. This happens about thirty seconds earlier at these locations than if they were aligned with the *Circle of Illumination*. Aligned with the *Circle of Illumination* are the Caracol in Belize (16.764°N 89.117°W), the Uaxactun Pyramid (17.391°N 89.629°W) in Guatemala and Calakmul (18.85°N 89.53°W) in México, the Medicine Wheel in Wyoming, US and in Alberta

Canada the Writing-on-Stone site and the Whakpa Chug'n Buffalo Jump sites -labeled "Maya Alignment" on the globe map. Note that these alignments are similar to the Candelabra's due to the proximity of the two locations.

In Cambodia the line passes by the Temples of Boeng Mealea, located forty miles east of Angkor Wat, Prasat Kho Ker, Prasat Khao Phra and the Holy Place at Wat Phu Lon; these alignments parallel the layout of the previous set.

Following The Mayan Great Arc line in its southerly direction, it crosses the *ecliptic* line. The approximate *ecliptic* line is shown in the graphic on p. 71. it connects Nasca and Cusco and continues across South America. Following the great circle drawn at this angle (68.67° NE), beyond past the archaeological site of Sacsayhuaman in Cusco, across the Atlantic in Egypt, the circle crosses over the Edfu Temple, built around 237BC over an earlier temple to Horus. Continuing on to Qatar, the circle finds the petroglyphs of Al Jassasiya, across the Persian and Oman Gulfs in India the circle passes over the Temples of Surya Mandir (Sun Temple) - Modhera, Suaminarayan, Tekari, Pandav Cave, Udayagiri, Kaali and Chandramauli. Across the Bay of Bengal in Burma finds the Golden Buddha in Ngapali and the Golden Rock, in Thailand finds the temples of: Prasat Prag Ban, Prasat Ku Pram and Prasat Yer.

SACSAYHUAMAN

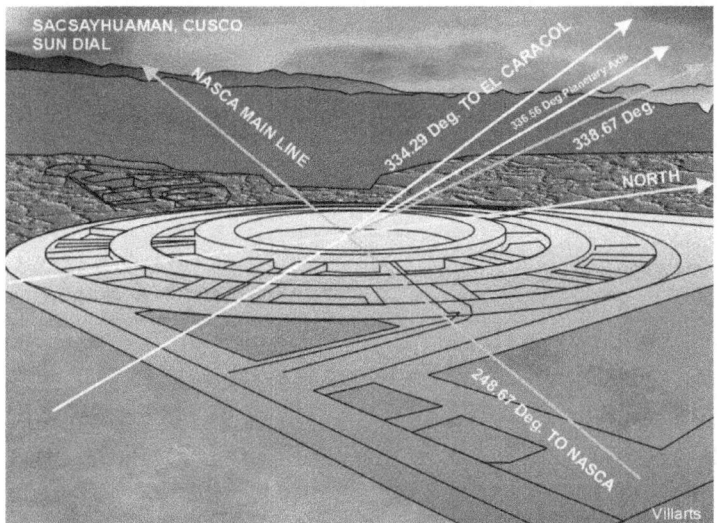

In Sacsayhuaman, known as the Inca 'Navel of the World', atop the citadel there is a circular structure in stone similar to the Nasca Calendar Wheel and the Medicine Wheel in Wyoming, US, known locally as "The Cusco Sun Dial", i.e. an observatory. See graphic above.

The Sun Dial has double row stone radii. One of the radius runs NW at an angle close to the *Circle of Illumination* and the other perpendicular to it i.e. parallel to the *ecliptic Equator* line. The 'dial' is another representation of the solar *Circle of Illumination* line; which at this latitude has an angle of 335.81°.The NW radius line (334.29° NW), connects The Cusco Sun Dial in Perú, with the Caracol observatory in the Yucatán peninsula in México, located 2,600 miles away.

The 'Sun dial' is another geodetic marker for the astronomical *Circle of Illumination*.

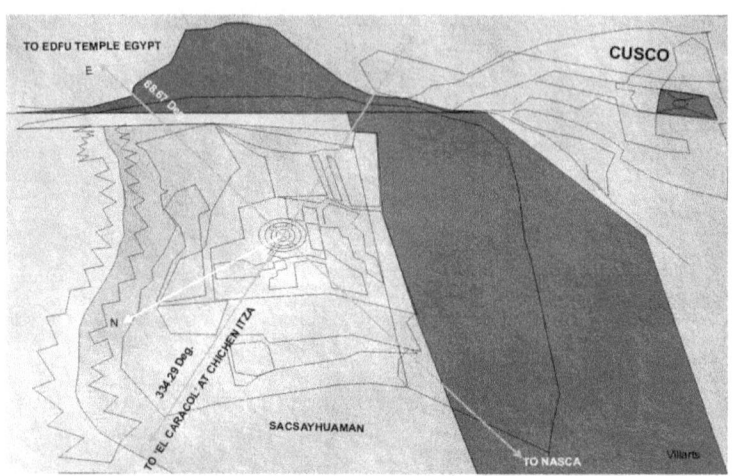

Sacsayhuaman is a site with a monument that encodes astronomical measurements in its geometry and its placement in a manner similar to the Candelabra, as was described: its location is found east on the 68.67° NE line that crosses the Nasca plain - the ecliptic Equator-, its Sun Dial structure encodes the angle of declination of the earth's axis and a line traced at that angle connects with the El Observatorio at Chichén Itza in Yucatan, México. In turn, the El Caracol Observatorio and the Castillo, both in Chichén Itza are aligned with the *Circle of Illumination,* at that latitude, at an angle of 25.17°. The data presents another piece of evidence that will help construct an argument as to why these sites were built, each at a specific location.

The alignments described and the structural features which display or point in the direction of the angle of declination of the earth's axis, begin to solidify the idea that the ancient architects had a plan; a plan that includes Nasca and Sacsayhuaman. To further add to the support of this idea and to delve into the true value of this discovery we need to look at Machu Picchu where we found the entire citadel was built around these concepts.

MACHU PICCHU

Machu Picchu is located 1.5 miles high in the Andes, northwest of Sacsayhuaman at about 302°. The citadel sits on a mountain ridge that runs northwest at 337.83°, coincidentally the ridge's direction almost equals the declination angle of the earth's axis. However, the Intihuatana Stone 'The Sun's hitching Post'-Another Sun Dial- lines up at 336.56°, the actual declination angle, with another structure atop the ridge. Also, the buildings' walls across the plaza and the terraced plaza below were constructed at the same azimuth, which points out the architects had knowledge of the *Circle of Illumination* which at this latitude has an angle of 335.89°. This would indicate that the site for Machu Picchu was likely selected because the mountain ridge follows the approximate declination angle of the earth's axis. The city architects planed the structures' alignment to correct the azimuth to more closely follow the declination angle of the earth's axis and also included the exact azimuth *of the Circle of Illumination's* angle in the stepped terrace's design.

The Sun Hitching Post stone's location (13.162°S 72.545°W) aligns with the sun at its zenith when it reaches the solstice culminations, at the spring and autumnal equinoxes, in the southern hemisphere, September 23rd and March21st. The value of the angle of the *Circle of Illumination* at Machu Picchu's latitude shows a discrepancy of 0.67° vs. the declination

angle of the earth's axis; the discrepancy is due to its location thirteen degrees latitude south. Because of this small angular difference the Intihuatana Stone or the Sun's Hitching Post casts a shadow at *clock time* noon at the summer solstice, however small it is; the sun angle is 4.38° degrees at noon. At *solar Time* noon the sun angle drops to 0.0°, thus casting no shadow[6], the sun's altitude is 77.07°.These metrics confirm the architects of this site had specific knowledge of planetary mechanics; this argument will be built upon as this discussion progresses.

In the graphic below the lines cross at the Intihuatana Stone. The faint gray lines follow the direction of the various structures that are aligned at 336.56°. In the photo the white line parallel to the gray lines shows the general direction the natural ridge follows.

The two circles at 336.56°and at 335.89° angle difference of .67° run together and in the Yucatan peninsula find the temples of Castillo near Tulum and Nonoch Mul in Cobá. The ellipses on their southern direction cross near the ruins of Chilenoyoc and further south Pachamarca. Near the city of Arica they find the Pukara San Lorenzo, south from there, in Chiza, an archaeological site with a number of anthropomorphic glyphs,

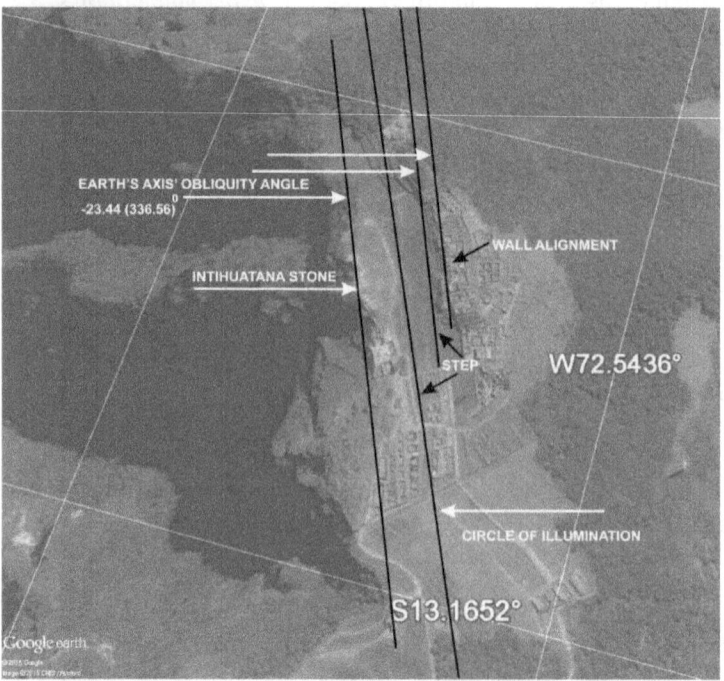

continuing fifty miles south is the famous Atacama Giant glyph and crossing into Argentina encounters the archaeological region of Pirichao. On their way to Indochina, in Indonesia one circle passes twenty one miles from the temple complex of Padang Roco and the other nine miles. Continuing the circles, in Viet Nam the circles find the Chiên Đàn trio Champa Temple, within three miles of both circles. All these sites are on the same *Circle of Illumination*.

A line drawn perpendicular to the gray lines at 66.56°- shown in white starting at the Intihuatana Stone- on its north easterly direction in northern Africa in Libya finds a standing stone Menhir in the desert and on the shores of the Nile in Egypt crosses by the Temples of Seti and Hator Dendra 26.141°N 32.672°E, located just north of the Luxor Temple. Further east in India it finds the Temples of Jain (24.35°N 70.75°E), Taranga, Birla Nagda and passes thirty miles south of the Stupas of Stadhara and Sanchi. Further east, in Thailand it finds the Phimai historical park and the Prasat Phanom Wan Temple and in the border with Cambodia passes by the Prasat Khao Phra, or Preah Vihear Temple, a large complex extending six tenths of a mile aligned on either side of a path with an azimuth of 182°, after this location Prasat Neak Buos is found. The sun follows a sequential path marked by all these structures at the summer solstice at each location, the solar angle reaches a maximum of 89.96° at Stadhara-Sanchi, then it declines to 39.74° at Machu Picchu.

Machu Picchu reaffirms the concept of *purposeful design;* its layout embeds the relationship between the circle of illumination angle and the earth's axis tilt. Machu Picchu's extraordinary architecture gives the site unparalleled importance, not only for all its archaeological attributes that have fascinated humanity for centuries, but for its *tie* with the Sun-Earth mechanics explicitly enshrined in the Intihuatana Stone and the layout of its buildings and plazas, in particular. The Nasca and Sacsayhuaman sites provide additional data points which support the concept of purposeful design. The illumination data for the three sites reveal a pattern previously unknown.

YURAK RUMI STONE

Is one of a group of precision carved megalithic stones found in the region; one is found in Machu Picchu; the Intihuatana Stone, the other in Saywite. The Saywite stone is equidistant to the Yurak Rumi and Intihuatana stones. The Yurak Rumi Stone is approximately 26 miles northeast of Machu Picchu. Its location is shown below. This boulder was discovered by Hiram Bingham on August 9, 1911[17]. The importance this boulder has is the geometric carvings found on it. The north side of the stone is cut flat facing northeast at an angle of 8.61°, the cut has an angle of 98.61°. A line with this angle is shown across the picture. The flat cut is slightly tilted from north, whereas the walls constructed surrounding the stone are aligned precisely north. The angle of the cut on the stone is emphasized by its design; the face is carved with a horizontal flute running lengthwise. The fluting and designs inside the flutes ensure the purpose of the cut at that azimuth; it discards the possibility of it being natural. Following the angle of the cut -the nearly horizontal white line in the picture- reaches Machu Picchu at the Intihuatana Stone; 'The Sun's hitching Post'.

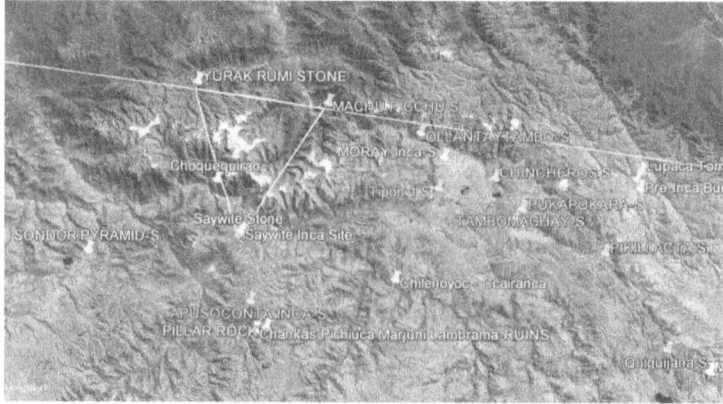

Above the flutes there are three protrusions carved on it, described by Bingham as serving the same purpose as the Intihuatana Stone, another ' Sun's hitching Post'. Continuing the arc at the angle of 98.61° over the Intihuatana Stone, in Sri Lanka crosses directly over the Sigiriya or Lion Stone, a 600ft high rock column where King Kasyapa erected his palace between 477-495CE[21], continuing east it passes 2.9 miles north of the Demala Mahaseya Stupa in Polonnaruwa, also atop a hill.

On top of the stone on the opposite side of the flat cut, there is an angular cutout. The angle lines which form the cutout run at 131.64° and 221.64°; a right angle. The first line runs southeast. In that direction encounters the following archaeological sites: Chilenoyoc, Sillustani and Puma Punku in South America. In Cambodia finds the citadel and temples of Phnom Chisor 11.185N° 104.823°E. Following the same line on its northwest direction at an angle of 311.64° the line finds the following archaeological sites in Perú: the ruins of Piruro, Canoncillo and Siete Techos. Across the Pacific Ocean to the north, in Japan the circle passes fifty miles north of the Yonaguni Pyramid and the megalithic Head in the Yonaguni-shima Island.

The Yurak Rumi stone is northeast of Nasca; a line connecting them at 232.96° aligns with a large parallelogram glyph found at 14.693° S 75.115°W on the plain. That is, its location was recorded at Nasca.

The alignments of the Yurak Rumi stone emphasize the concept of viewing ancient structures as geodetic markers. The Yurak Rumi Stone points to Machu Picchu in a manner similar to the way the Candelabra points to Nasca. This method, we found, is not limited to any particular type of structure it can also be found in the geographical alignment of menhirs arranged to line up with sites thousands of miles away as is the case with this rock and others that will be discussed in the course of this study. Two examples of this, we visited earlier, are the ring structure's alignments that encode an astronomical measurement, mentioned before: the Medicine Wheel at Big Horn WY, US (44.83°N107.92°W) and the Calendar wheel at Nasca, (14.64°S 75.17°W) which are about 4,500 miles apart. The line that connects their centers has a heading of 335.27°, which as was pointed out before this angle is close in value to the declination angle of the earth's axis of 336.56°; this line parallels the Sacsayhuaman Sun Dial to El Caracol line (334.29°) which also has similar angle to the *Circle of illumination.*

It is worth noting that the Sun Dial/Caracol and the Medicine Wheel/Nasca Calendar lines are found on either side of the earth's axis line at Nasca -The Mayan Arc. Both lines connect sites that are virtually identical in design and each pair is located on the same side of the axis line: The western line connects the 'Wheels', while the eastern line connects the 'Observatories'.

The Medicine Wheel/Nasca Calendar line passes over The Cueva de los Tayos in Ecuador, a site of controversial archaeological value explored in the '60s', which gains in significance as this study reveals, and which will be discussed later.

Recapping; there are four archaeological sites in this region with design elements which demonstrate that those who created them knew about the declination angle of the earth's axis. Consequently the *Circle of Illumination* was also known to them as demonstrated by the alignment of pyramids and 'sun dials' along the circle to receive sun light simultaneously. This theme repeats around the globe and as will be discussed later, these arrangements create a global pattern; a pattern with precise astronomical significance.

From the foregoing analysis another pattern emerged that was unknown until now: The Nasca lines are part of a directional map which indicates where other archaeological sites are found. That is to say; the lines at Nasca are arc segments of great circles (ellipses), in the direction of which monuments were built and, perhaps, in some instances became places where our current civilization flourished in ancient times. The temples found in Nasca's antipode region in Cambodia and Thailand, each one of them has its corresponding line engraved on the Nasca plateau and for some of them their circles pass through Machu Picchu.

To test these results, new connections were made in reverse. Several temples located in the antipode region were selected and marked. Lines were extended from the markers all the way to the Nasca plateau; it was confirmed, each line could be fit to a line on the plateau.

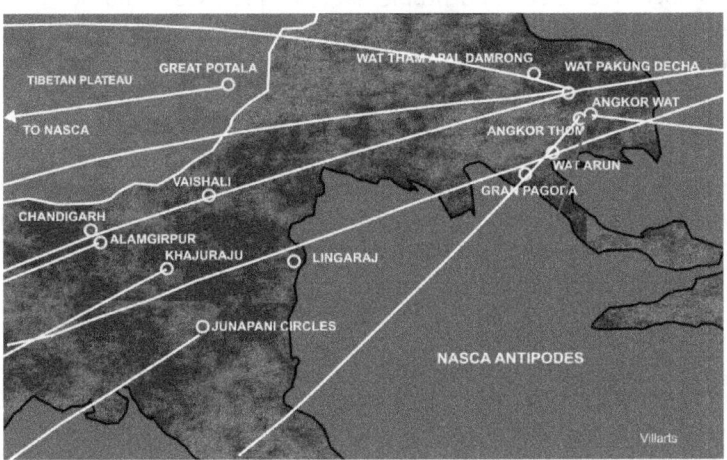

We extended the test to other archaeological sites and ran lines to Nasca; it was discovered that not only temples have corresponding lines at Nasca, but virtually every site of archaeological interest we tested has a line, as well. The graphic above shows how the lines look like for a few of the temples in Nasca's antipode region.

Further study showed that not only pyramids, dolmens, temples, circles, cairns and any other kind of site of archaeological interest has a corresponding line in the Nasca plateau; so did all anthropological sites we had included in the first test. In the graphic below one can see Taung Child and Sterkfountaine's line at the bottom of the graphic.

Other sites from North America are also recorded with their own lines: Hovenweep, UT and the Medicine wheel at Big Horn, WY. Also, archaeological sites in India; Alamgirpur and Valishali are shown. The pyramids of Cheops and Chephren, and the Sphinx appear towards the top center. Also shown are Chichén Itza, the Golan Circle and El Infiernito.

The logic behind the mingling of archaeological and anthropological sites seemed problematic. At first, we could not imagine why the designers of the lines, would want to record the locations of our ancestors, to consider it important enough to mark the locations they were found in. Why would 'explores' want to do that?

> When the data was reviewed as a whole, a far-flung reason was found that could explain it: It would support the same rationale we had used to include anthropological sites in the first part of the study; Mr. Sitchin's assertion regarding the alien's need for a labor force, predating human existence. These data appears to support that theory; the sites would be the sources for hominids or prehistoric man.

The archaeological site locations and their corresponding Nasca lines were marked and cataloged according to, country location, type, coordinates, their distance to Nasca, the length of the corresponding Nasca line and its bearing.

A sampling of the data is shown on the next page.

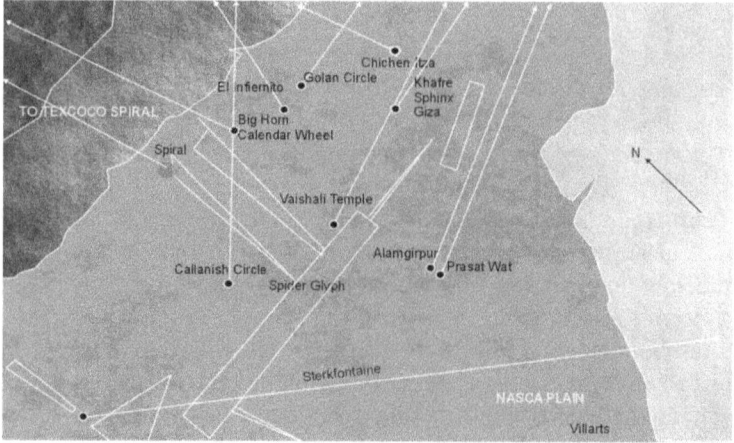

SITE	Country	Type	Latitude	Longitude	Miles to Nasca	Nazca Line lgth miles	Angle
Infiernito-Villa de Leyva	Colombia	Alignment	5.65	-73.53	1402	0.09	4.66
Poverty Point Louisiana	USA	Observatory	32.64	-91.41	3428	0.44	17.95
Valishali Temple	India	Temple	26.00	85.08	10948	0.21	27.35
Brogard circle,Orkney Islands	Ireland	Observatory	59.04	-3.15	6465	0.29	29.34
Callanish Circle	Lewis	Observatory	58.18	-6.73	6319	0.09	30.50
Caral-Major -Minor	Peru	Pyramid	-10.89	-77.52	305	0.29	30.50
Neanderthal Man- Erkath	Germany	Site	51.22	-6.92	6664	0.08	39.23
Stonehenge	UK	Observatory	51.18	-1.83	6380	0.87	42.05
Altamira Cave	Spain	Site	43.38	4.12	9647	0.10	42.90
Lascaux Cave	France	Site	45.05	-1.18	6284	0.12	43.39
Neanderthalis-Le Chapelle-aux-Saints	France	Site	44.99	-1.73	6305	0.11	44.06
Chauvet Cave	France	Site	44.39	-4.41	6420	0.11	45.00
Sulemayan-Costantinopla	Turkey	Site	14.71	75.22	7616	0.05	51.16
Homo erectus ergaster georgicus	Georgia	Site	41.33	-44.18	8399		51.24
Great Potala	Tibet	Temple	29.65	91.12	11089	1.32	51.70
Wat Pakun	Thailand	Temple	15.98	103.48	12430	1.92	53.40
Gobekli Tepe	Turkey	Observatory	37.22	38.92	8145	0.10	56.02
Prasat Wat Pho Siri That	Thailand	Temple	15.30	103.93	12368	0.27	56.90
Machu-Pichu	Peru	Citadel	-13.16	-72.55	202	0.04	58.00
Atlit Yam stone circle submarine	Israel	Observatory	32.71	34.94	7919	0.07	60.90
Rujm el-Hiri – Gilgal Refaim-Golan Circle	Israel	Observatory	32.90	35.80	7971	0.05	61.10
Alexandria -Rhakotis	Egypt	Site	31.20	-29.90	7,620	0.03	62.12
Jerusalem	Israel	Monolith	31.77	35.21	7937	0.64	62.25
Saqqara Djoser Step Pyramid	Egypt	Pyramid	29.87	31.22	7688	0.17	63.25
Khaujaro Temples	India	Temple	24.86	-79.92	10680	0.10	63.83
Khafre	Egypt	Pyramid	29.98	31.13	7684	0.51	63.86
Moraya Inca	Peru	Site	13.32	-72.18	216.25	0.00	63.95
Great pyramid Giza	Egypt	Pyramid	29.98	31.13	7683	0.13	64.12
Sphinx	Egypt	Monolith	29.98	31.14	7684	0.51	64.14
Newgrange	Ireland	Mound	53.69	-6.48	6196	0.04	64.56
Monsaraz Cromeleque Stones	Portugal	Observatory	38.56	8.06	5702	0.07	64.58
Basrah	Iraq	Site	30.48	47.82	8680	0.15	64.78
Carin of Bernenez	France	Tumulus	48.67	3.86	6152	0.11	66.37
Alamgir Pur	India	Site	29.00	77.48	10,436	0.33	66.38
Nabta Playa Circles	Egypt	Observatory	22.53	30.70	7603	0.26	66.92
Carnac Stones	France	Alignment	47.58	3.08	6155	0.12	66.95
Castlerigg stone circle	UK	Observatory	54.60	3.10	6350	0.06	67.03
Tumulus du Rocher	France	Tumulus	47.63	2.95	6163	0.61	67.06
Mambury Circles	UK	Observatory	50.71	2.44	6267	0.32	67.45
Pre-Inca Burial	Peru	Site	-13.42	-71.36	252		69.06
Tumulus of Bougon	France	Tumulus	46.37	-0.06	6258	0.09	69.17
Junapani Circles	India	Observatory	21.00	79.00	10692	0.13	72.46
Luxor at Karnak	Egypt	Citadel	26.07	32.09	11134	0.10	76.07
Tarxien Temples	Malta	Temple	35.87	14.51	6789	0.22	77.81
Ari Ramidus	Ethiopia	Site	10.48	-40.45	8110	0.29	77.89
Australopithecus	Ethiopia	Site	10.28	-40.53	8114	0.04	77.89
Kasberga Stones	Sweden	Alignment	55.38	14.06	7012	0.02	80.86
Baalbek	Lebanon	Monolith	35.14	38.89	7997	0.59	82.97
Homo Sapiens	Ethiopia	Site	10.43	-40.57	8108	0.06	85.80
Rajajeel	So. Arabia	Monolith	29.89	40.22	8232	0.65	90.95
Latte Stone Site	Guam	Monolith	13.33	144.68	9776		96.93
Puma Punku	Bolivia	Citadel	-16.55	68.67	451	0.31	106.42
Tihuanaco-Kalasaya	Bolivia	Citadel	-16.55	-68.68	451	1.31	106.54
Sterkfontaine-Taung-Cradel of Humanity	So. Africa	Site	-26.02	27.16	6555	1.03	119.00

During the line-matching process it was realized that the Nasca lines differ in condition and or design. Some are sharp and easy to follow where others are 'worn-looking', while others are a composite or re-drawn. Some appear in relief, or marked with stones and many are so faint they can be seen only at low altitudes. Furthermore, some of the lines belong to 'sets': a triangle, a strip with parallel lines or an elongated parallelogram, each of these varying in prominence and size. There may be a correlation between the reported age of a site and the apparent age of its corresponding line. The hypothesis is that these lines were created over the course of thousands of years. Therefore, if a correlation were to exist, the apparent age of the line may coincide with the age of the site it points to. For example, the lines for Homo sapiens, Peking Man, Neanderthal and the Altamira Cave all are part of trapezoidal shapes of different sizes -and ages? Tung Child & Sterkfontaine is a wide line, apparently re-drawn three times; while Chauvet, Lescaux and Java man are marked with thin faint lines.

It is noteworthy, that according to some accounts there are from 300-800 lines on the Nasca Plateau. Of the hundred or so lines we followed to archaeological sites, except for a few, most of the large obvious lines or trapezoids did not to lead anywhere; the few that did belong to prominent places, i.e. Easter Island and The Athenian Acropolis.

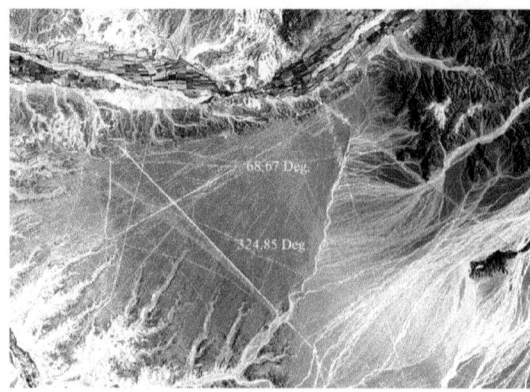

Nasca lines crossing, 324.85°NW and 68.67°NE. NASA Photo.

The majority of the lines that were recorded were short and 'hard to find'. The large 'landing-strip-like' lines were the first evaluated, with no apparent result. We conjecture some of the unmatched lines do lead to places not yet identified. The first alignment found of unprecedented significance was that of Jupiter's Temple at Baalbeck; its line at Nasca is one of the longest it measures 0.59 miles. Its significance was emphasized by the fact that the line passes over the Sondor Pyramids (13.6°S 73.26°W) and atop Machu Picchu (13.16°S 72.54°W) in Perú- and in northern Africa in Libya, goes past the ruins of 7th Century BC Ptolemais (32.705°N 20.95°E).

Equally surprising was the discovery that the Athenian Acropolis has a line at Nasca. The line reaches Nasca from the Acropolis at an angle of 263.54°. This line passes over the Yurak Rumi Stone. The Acropolis is not aligned with the cardinal points; its east to west alignment angle is about 257.0°. On the plain the 263.54° line finds a prominent double line that connects two prominent Nasca quadrangular glyphs, shown on the graph below. The top line of the quadrangle on the left of the graphic aligns with Easter Island; this establishes a geometric connection between Easter Island with the Acropolis.

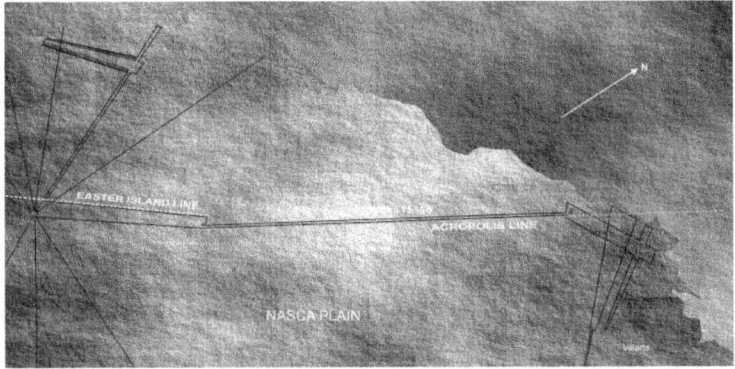

The alignments of Nasca with Baalbek prompted a closer look at Machu Picchu's alignments. Running the line that connects Sacsayhuaman and Machu Picchu (at 302.21°) on its southerly direction crosses the ruins of Iskanwaya (15.47°S 68.68°W), further south it finds the petro glyphs of Kellkakta. Some of the glyphs at this location (15.0°S 67.83°W) are virtually identical to the Sonora Stone's: Spirals and a Lemniscates of the bow tie type. Continuing south the line crosses eighteen miles east of Inkallakta (17.6°S 65.42°W). The wall ruins that remain at this location have the same architecture as Machu Picchu's; not surprising since it was part of the Inca Empire. Following the line further south, before crossing from Bolivia into Brasil the line cuts through the Samaipata Fort (18.18°S 63.8°W).

VERIFICATION

To test the alignments for randomness or simple coincidence, a 'Nasca line map' consisting of a number of lines having a one-to-one relationship with their counterparts, in size and direction in Nasca, was drawn and 'transported' to the Salt Flats at Bonneville Utah. No sites could be matched from there. However, one could create another 'Nasca in reverse', i.e. draw lines from any archaeological site and take them to any desired remote site and mark lines; thus reproducing something similar to the original Nasca. That experiment, to some degree, would show that Nasca could have been located anywhere at random, except, there are constraints that determine its location; the Nasca lines are geometrically tied to the Candelabra's design. Several of the sites predicted by the Candelabra, as previously shown, are also mapped into Nasca lines. But, those are not the only constraints. There is another 'Hub', a third location, geometrically tied with Nasca which we think was the original map: Easter Island.

The island is also known as Rapa Nui and Isla de Pascua. At this location the Island's Ahus and Moais determine the positioning of the circles (ellipses) which tie together these three archaeological sites and also map other archaeological sites around the world.
 A map showing these three 'Hubs' is shown below.

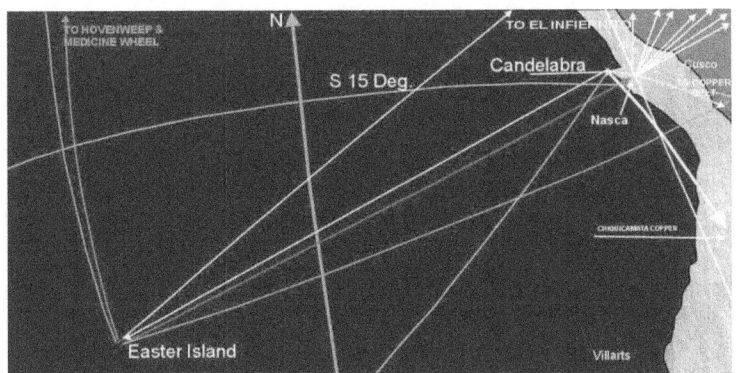

With the discovery of these alignments, we conjectured, that to design and execute such a great world-spanning plan, it would have been necessary to develop a local 'project control' method, an 'instrument' that could have been used to keep track and maintain a record of the location of, possibly, hundreds if not thousands of sites; some of which we have catalogued: Table p.87. The site needed to be invariable over millennia, virtually impervious to human, animal or atmospheric disturbances.

> If extra terrestrials were involved, their visits, they knew, would take place over thousands of years. As, Mr. Sitchin explains, it would take their planet Nibiru's orbit 3600 years to bring the aliens back within reach of the earth[2].

With the results of this research in hand, we realized we were looking at the place that met the technical requirements just described: the Nasca plateau is such a site.

It is located in one of the driest places on earth, just north of the Atacama Desert area. It has no water or overwhelming amounts of sand that could obliterate its existence.

This location is literally rock-hard and impervious to damage, even today in spite of the modern tourist who has marred its surfaces with tire tracks and even graffiti. The attack has be so relentless that person's names are carved on the rocks nearby and in Arica, another archaeological site to the south of Nasca, there is a gigantic Coca-Cola billboard carved on the rock on the side of a mountain.

The glyphs and lines are but a few inches deep and have survived the test of time. The site could have been a landing site, as claimed by many of the proponents of alien visitations, but, in our estimation that would have been just incidental to its purpose. The Nasca lines served as a cartographer's survey map over the ages; we think that makes more sense and this study provides abundant data that supports that theory.

The map on the previous page shows the three sites with some of the lines extending from them, the layout of which reveals a fact about them. On the graphic, we can see that the lines emanating from each central location, on their way to remote sites, sweep the horizon. The location and bearing of the Nasca and Candelabra lines and the alignment of the Ahus' and Moais' appears to follow what could have have been a systematic plan.

The directional angle or azimuth of each line identified at Nasca was measured and cataloged and the data sorted from the smallest to largest angle; the smallest angle is close to 0° geographical north. The first dark line at the top corresponds to El Infirnito at 4.53° on the pie chart plot shown below. The graphic data plot visually reveals the plan or design; the Nasca lines' bearings are distributed over 360° in a systematic manner. This fact points out something that was not obvious at the onset. What originally appeared to be a haphazard scattering of lines, we had referred to as a child's drawing; it turned out the lines are geometrically arranged.

The systematic sweep shows that every point on earth, potentially, could have a line at Nasca, however, it doesn't seem to be the case. The plot of the angle data -graph on the next page- shows that the lines concentrate pointing towards the NE quadrant. From the data we can establish there is a relationship between each site and a Nasca line, however it is not possible to determine which was created first. We assume the sites' locations were determined before the 'maps' were created.

We believe the ancient travelers left markers at each site they visited.

They constructed or guided the construction to suit the need at the time; from simple menhirs used as route landmarks to sophisticated constructions to record astronomical events and the earth's planetary mechanics. Many of these constructions may have been re-built or built upon, layer over layer generation after generation. Their original purpose obliterated through legend and lore; more often than not giving them a mystical religious coloration, which through the ages evolved into a new reality paradigm.

Thus far we have discussed five archaeological sites, two of them are cities: Machu Picchu and Sacsayhuaman, the other three sites; the Easter Island's Ahus and Moais, the Nasca lines and the Candelabra, are uninhabited mapping executions. These five interconnected centers, together, change our view of the ancient travelers from being a primitive localized civilization to a technically sophisticated civilization that had a world view. But, this doesn't end here. We will show this mapping technology unfolds to encompass many archaeological locations; from simple menhirs or cairn markers to well developed 'hubs' such as these first five and ultimately a coherent global plan.

The data distribution also shows it would have not been possible to create the 360° radial arrangement of all the lines found in Nasca, following the rise and fall movements of planets from this geographic latitude of about 15°. All other theories, presented until now, seem to fade as well.

However the cultural anthropology of each location may not necessarily be disturbed by the existence of an ancient traveling race, at least not in its totality.

We begin to see there may have been a layering of cultures through time. The radial arrangement of routes we discovered is not limited to these locations; others will be discussed in the following chapters.

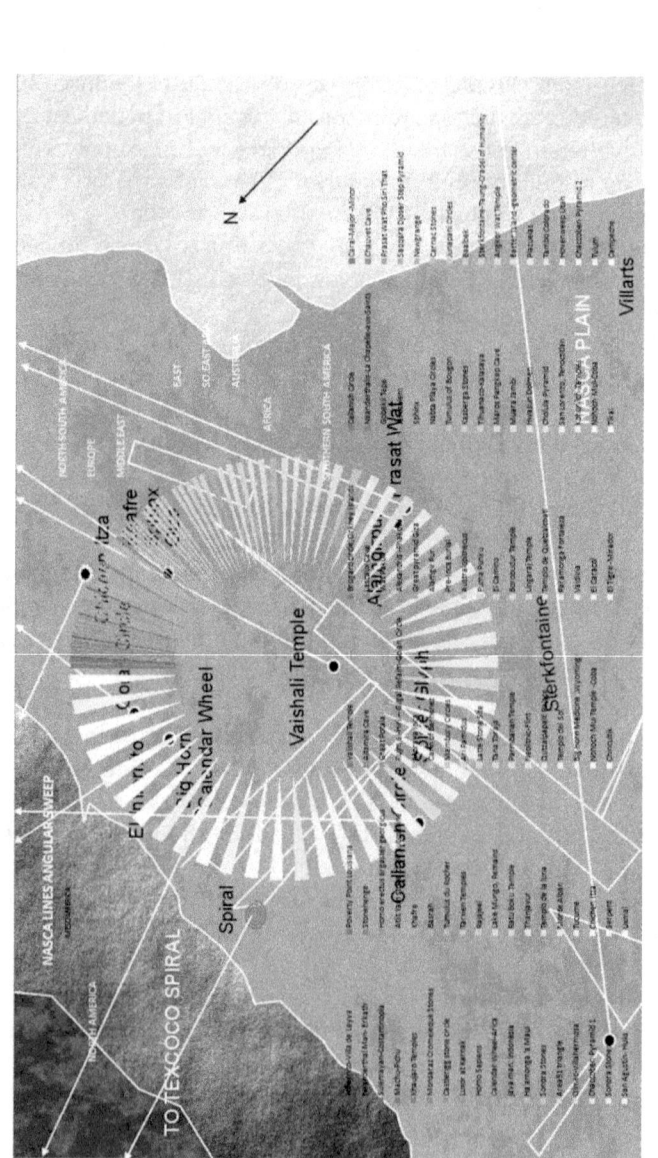

WERE THEY ALIENS? Since the data shows a systematic sweep of the horizon, it could be argued that chance alone is responsible for some Nasca lines to point to the locations where early man was found. However, if we recall Mr. Sitchin's narrative, it provides a justification for the purpose of marking early man's locations. The aliens, he claimed came to earth to mine minerals. To develop manpower, they created man to do the mining labor, by crossing their DNA with the hominids'; which he supports with the Bible line "God created man to his own image".

Suspending judgment for a moment and putting together our results and that claim, it would follow that when the aliens visited earth they could have searched for help to carry out the mining operations but all they found were hominids who were transformed by genetically engineering man and to ensure the supply they marked every place where they were found. The first generation humans, he adds, became the half-Gods. If the aliens engineered demigods everywhere they were found, then Eden would have been located in multiple places around the globe and the extraterrestrial DNA would have been mixed with different hominid species; which could explain the difference in today's human races. It is said: we were all created equal, but given this argument it should be said: equally, via the same method? The concept of multiple Eden locations may be the basis for the various legends each culture has about creation, which share some commonality in their descriptions, such as: "que en las regiones cercanas a Tunja existía la Laguna de Iguaque, de cuyas aguas emergió Bachué, nimbada de una luz que hizo resplandecer la tierra…". "…near the Tunja region the Iguaque lake existed, from which the Godess Bachué came out of its waters holding the first human child in her arms, drenched in a bright light that illuminated earth ."[61]

That is the narrative of the legend of Creation of the Andean Chibcha Tribe civilization in central Colombia. The image this legend presents has similar elements as the biblical Annunciation, although in this case the interaction between the light with a female from earth resulted in the creation of man, not mankind's Savior.

New DNA ancestry tracking technology will shade more light in this regard. The most recent research tends to show that all of mankind originated in Africa about 4 million years ago and started migration about a hundred thousand years ago. For current research on this topic visit: http://www.hhmi.org/biointeractive/using-dna-trace-human-migration

Notes

CHAPTER 5

EASTER ISLAND-
ISLA DE PASCUA - RAPA NUI

"What we are suggesting therefore is that Easter Island might have originally have been settled in order to serve as a sort of geodetic beacon, or marker – fulfilling some as yet un guessed at function in an ancient global system of sky-ground co-ordinates that linked many so-called 'world navels". Graham Hancock-

Easter Island is located in the Pacific Ocean in the furthest point from land than any other place on earth.   A line drawn westward from the Temple on the Mount in Jerusalem, at 270°, which we had earlier established its dome is aligned EW with the Mount of Olives and the Pater Noster Cathedral;

reaches the Island seventeen miles off its north shore, the island is 12,351 miles away, exactly half way around the world

A line drawn west at 270° from the Golan Circle passes the island fifty three miles south of it; while a line that connects the Golan circle with the Hanga Te'e circle in the island has a bearing of 271.5°. The 1.5° difference in angle appears to indicate, the locations of both, the Temple on the Mount and the Golan Circle are determined by the position of the island. The graphic on page 42 shows the location of Jerusalem and the Golan circle.

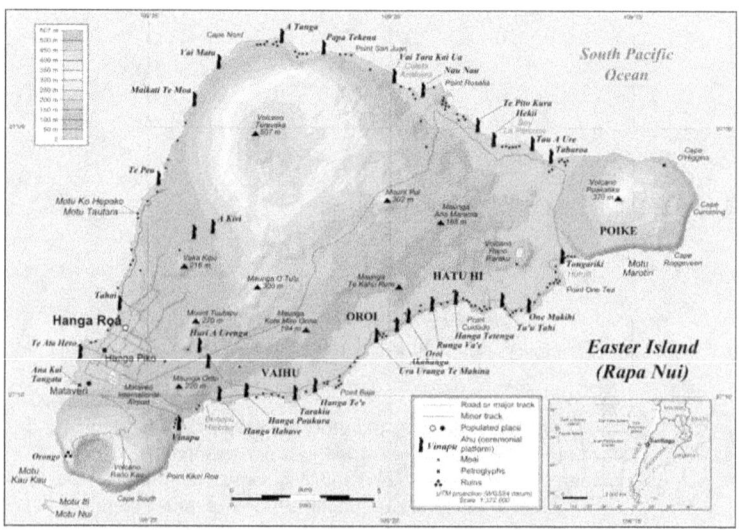

Easter Island Ahus and Moai Locations - Wikipedia Creative Commons

Many of those who have wondered about Easter Island's mysteries, looking at the Moais have asked: What are they looking at? We discovered that the only way to know would be to follow their gaze. The Ahus with their Moais are located around the island as shown in the map above. Surveying the Ahus, the pedestals where the Moais stand, it was noted they are not arranged in any particular manner.

That in itself is a clue. This clue reminds us of the messy splatter of lines at Nasca which turned out were geometrically arranged in a unique way. The one place in the Island, where this is obvious is north of Hanga Roa Beach at the Tahai complex (27.14°S 109.43°W).

There are three Ahus at this location: Ahu Vai Uri, Ahu Thai and the famous Ahu Ko Te Riku moai; the one with eyes. This is the moai who probably inspired the famous question: "What are you looking at"

To find their alignments, lines were drawn lengthwise across the Ahus and their bearings measured. It was found, all Ahus are aligned in a different geographical direction. Therefore, the Moais atop each Ahu look in different directions. See graphic below.

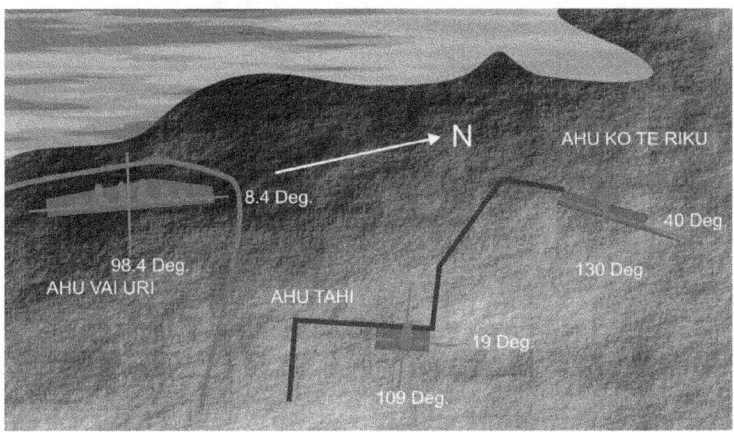

Ahu Vai Uri, Ahu Thai Hanga Kio E and Ahu Ko Te Riku Tahai

For the study of Ahu alignments, we applied the same 'Follow-the-line' technique that was applied to the Nasca lines study, but with a variation. Lines were drawn following the Ahus direction and a second line was drawn, for each, following the line of sight for the individual Moai; that is, perpendicular to their line-up or the Ahu itself, front and back. Some Ahu platforms are trapezoidal and stacked in a pyramidal manner, resulting in the layers having different azimuths. Ahu Ko Te Riku is a prime example; the small azimuth differences between the layers point to specific sites around the globe.

Some Ahus have lost their Moais and some are in a poor state of disrepair and the lines can be drawn only in their general direction. Also, there are various levels of disrepair. Some are visible as a trace or a scatter of toppled Moais and their Pukaus. As with the faint lines at Nasca, in some instances, it was difficult to assess their bearings accurately; so, several readings were needed to confirm the results. It should be noted that several of these sites have been repaired, so we have assumed their original alignments were kept as close to the originals as possible.

At first we had considered the island as a location, as a single archaeological site. Therefore we proceeded to find out if the island had a line on the Nasca plain as the previous sites that were tested did. We found the island's geometric center to set the marker for the connecting line, which served to find the island's line at Nasca (The Island is an almost perfect isosceles triangle so it was easy to find its center). See graphic on page 98.

After determining the connection between the two sites, lines were extended from several of the Moais into Nasca. Again, we found each of the Moais had a line. Lines were found for; Ahu Vinapu, Hanga Te'E, Hanga Pokura, Kote Riku, Te Peu and Tahi. These lines were not necessarily in line with or perpendicular to the ahus.

Applying the Ahu alignment technique previously described, we proceeded to draw circle-ellipses extending them around the world to find out where they lead. One more time surprising results were obtained; a series of connections were made to archaeological sites, fewer in numbers as compared to Nasca, but more extensive than the Candelabra's. The Candelabra is also connected to Easter Island in this manner. Ahu Tetenga's line 72°NE direction crosses atop the Candelabra's right and left first lobes.

Some of the Ahus, with or without Moais -most of which are found in the island's perimeter, their structures point to archaeological sites of interest.

A few of these remote sites have geographical alignments of their own which map out directions to other locations at various angles, these sites are 'hubs' in themselves. The El Infiernito archaeological site in Villa de Leyva, Colombia and Göbekli Tepe, both mentioned earlier in Chapter 1 and the Chaco and the Hovenweep Canyon Pueblos, in the US, are some of these sites.

Here we pause with the Easter Island analysis to discuss the archaeological sites of El Infiernito and Göbekli Tepe, both are 'hubs' and the relationship they have with other sites in the region and globally. The Chaco Canyon and Hovenweep Pueblos sites are discussed in Chapter 11

EL INFIERNITO

The El Infiernito (5.65°N 73.56°W) site is connected to, both, Easter Island and Nasca. The site consists of a central phallic megalith surrounded by a circular arrangement of similar megaliths -a graphic of the layout of the site is shown towards the end of this discussion, p.109.

Ahu Tu'U Tahai (27.13°S 109.30°) has bearing of 51.89°. A line drawn following this azimuth points to the phallic Megalith Giant; continuing the line north past the megalith, to the right of it there is a burial dolmen found at the site a few feet away from it.

The phallic megaliths at El Infiernito are not unique in design, the largest megalith known with this design is found at 40.67°N 106.33°E west of Beijing in China, which is located exactly due north of the El Infiernito's central megalith 9,253 miles away.

On its way to China, in Ireland the line passes 24 miles south of the Tara Stone and over the Rudston phallic megalith in the UK (54.09°N 0.32°E).

The El Infiernito site has twenty four standing, phallic megaliths and several fallen ones; twenty three are arranged around the middle one and the largest, at specific angles starting from the northwest at 291° ending on the southwest at 175.59°. Included are one at 336.0°, close to the angle value of the declination angle of the earth's axis and one at 89.68°, close to the celestial Equator's angle. The Ahu Tu'U Tahi line to El Infiernito, connects the two locations through the region where the Valdivia culture flourished between 3500-1800BC, along the shores of South America, from La Rinconada to Ayangue, in western Ecuador. The distance between the two end points of the line is 3,294 miles.

The Phallic Giant-Wikipedia Commons Photo

The El Infiernito site has a Nasca line, found at 14.68°S 75.11°W. Following the line from Nasca, north past the Megalith Giant, it reaches the Sierra Nevada de Santa Marta in northern Colombia on its eastern side. This snow covered peak is unique in that it is located about forty miles from the Caribbean seashore.

Between the peak and the shore, is the site where the Tyrona Civilization flourished, estimated 4000BCE. The so called 'Lost City of Tyrona' is found thirty miles from the shore. The El Infiernito line passes sixty miles east of it. We'll discuss the importance this has a bit later. The El Infiernito site was determined to have particular significance, although it is different from most other sites in the Americas. The site's megalith alignments in a radial configuration, gives this archaeological site importance equal to that of Stonehenge, Avebury and Mystery Hill in New Hampshire, also known as America's Stonehenge and several others like the rings of; Brodgar, Callanish and Templewood at Kilmartin to name a few.

This arrangement of megaliths and dolmens with a prominent giant monolith amongst menhir alignments is not unique. A similar arrangement is found at Carnac in southwest France, near the Manio Quadrangle. East of the quadrangle is the monolith Manio Giant (47.6°N 3.05°E), together with several dolmens found nearby.

El infiernito menhir alignments

The area where the Carnac menhir alignments are found is much larger than El Infiernito's. It is about 1.5 by 2.5 miles and the number of alignments count in the dozens.

Earlier, we had stated that alignments not only follow astronomical events, but sometimes, their arrangements within a site may point in directions that may align with other important sites. At the El Infiernito site, there is also a menhir alignment (see picture above) with a 90° bearing, located just west of the central megalith. The alternate purpose of the meter high megalith (menhir) row appears to be geographic; it sets the coordinates more precisely. Astronomically, the menhirs cast different shadows at various times of the year, thus marking the solstices. In September the alignment appears to point directly to the sunrise in the east. On September 23, 2014 at 6:06am sunrise the solar azimuth was 90.26°. At solar time noon it was 180°, altitude 84.12°. A few steps to the south of the alignment there is a 'circular' menhir arrangement with eight large and several small stones.

Spring Equinox

El infiernito Dolmen-Wikimedia Commons

The center of this 'circle' is at about 250° from the Megalith Giant. A line drawn in this direction points to the site where Java Man (Pithecanthropus erectus) was found in Indonesia. On the line's path forty nine miles east of Tahiti is where the Tevaifaara phallic megalith is located (17.57°S 149.32°W), further south, in the Island of Tonga, it passes by the Ha'monga Trilithon, p.127 (21.14°S 174.05°W) and in northern Austarlia finds the Litchfield region where Copper, Tin and Uranium are mined.

A line traced connecting the Megalith Giant with another megalith found at 317.73° cuts through the Yucatan peninsula near Chichen-Itza and in Arizona, US goes through the Chelly Canyon Pueblos. Another alignment line at 321.88°, points to the Niah Cave in Malaysia. This archaeological cave is the site where human remains were found dated 40,000 years BCE. On its way, it goes over Salt Lake and the Bingham Canyon Copper Mine in Utah and in Oregon goes over the Columbia River petro glyphs near Cascade Island and further north by Pillar Rock.

The line at 305° confirms the center megalith's alignment with the Dolmen at the site at 35.0°, exactly 90° from it. The dolmen is an underground burial site, in its interior there are megaliths aligned in two rows that point to the east. See picture in the previous page.

Continuing the 35° line in its SW direction, it points to four important archaeological sites within Colombia. Immediate to the south is the salt mine in Zipaquirá (3.02°N 74.04°W).

Salt Cathedral, Zipaquirá, Colombia-Wikimedia Commons photos

This mine is dated probably early classic period, although the formation is dated 200 million years old. In 1954 a cathedral was built in one of the main excavated caverns. It was closed for safety reasons in 1990, and was replaced in 1995 with a more modern, architecturally planned one, shown above. The cross is about 165 feet high[18].

Further south on the line, is the El Abra Cave a burial site (4.59°N 74.3°W) located at 8,157 feet above sea level, it is a cave system dated late Pleistocene. In the cave there are numerous glyphs, one is an anthropomorphic glyph that has four hands that display a similar design to others mentioned before. Not only are the hands similar, but the object next to it is reminiscent of an upside down candelabra.

South from this location are the archaeological sites of Tierradentro (2.58°N 76.03°W) where carved megalith stelae are found and the National archaeological park of San Agustin located at(1.9°N 76.29°W). San Agustin is a UNESCO site which extends over an area of over one hundred hectares and comprises 600 known megalithic statues and forty burial mounds dated between 1-900AD[40]. The San Agustin Archaeological Park is, also, renowned for its carved megaliths; the necropolis is shown below.

Tierradentro & Necropolis- Wikimedia Commons

Now following the line in the 35°NE direction, it goes over the Chicamocha River Canyon where numerous geo glyphs are found at a site known as "Mesa de los Santos". One of these glyphs is strikingly similar to the Medicine wheel or the Nasca Calendar and, also, to petro glyphs found in Beleza São Desidério, Brazil and the Chaco and Chelly Canyons. Glyphs with this 'sun' pattern are believed to depict a high energy event from space[27].

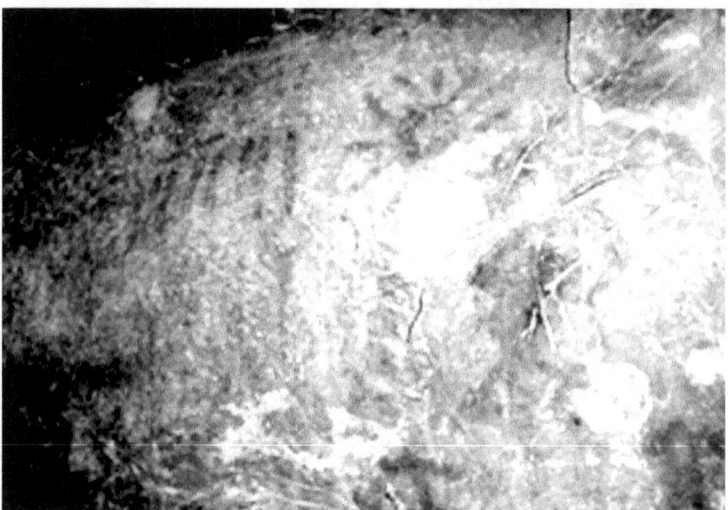

Mesa de los Santos- Chicamocha, Colombia Wikimedia Commons Photo

Continuing the line north across the Atlantic, in Ireland it finds the Newgrange Mound, it also passes five miles north of the Hill of Tara where the famous 'stone of Destiny' stands. The Lia Fáil stone, as it is called, is also a phallic megalith. Further north encounters the Egton Moor megaliths in the UK. Following the line over the north pole, now in a southerly direction, in Kazakstan, the Utzyurt Ruin megaliths are found, then onto India, the line crosses the Baijnath Shiva Temple (32.05°N 76.65°E).

In Indonesia the line passes 200 miles west of the location where Java Man was found (7.4°S 110.82E). The Palau Islands in Indonesia is the antipode region to El Infiernito. El Infiernito is on the same circle of illumination at Gunung Padang a similar megalithic site. In northwest Australia it finds the salt flats near Balla Balla, the Christmas creek iron mine and the Newman petro glyphs, before returning to Colombia it passes near the Chimborazo Volcano in Ecuador.

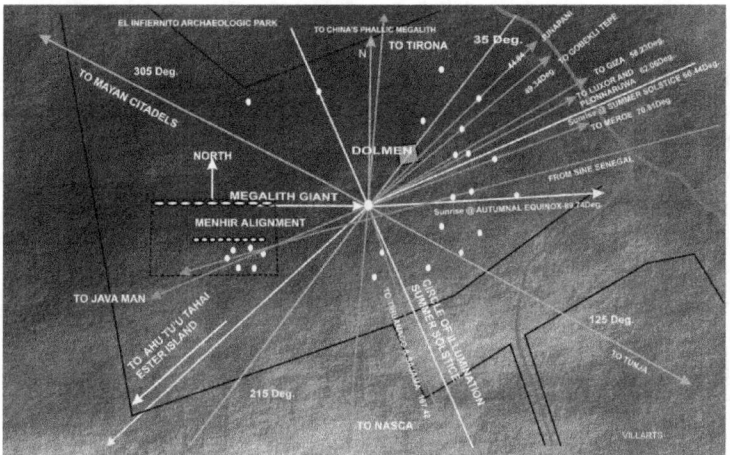

El Infiernito, layout drawn to scale.

A line drawn perpendicular to the 35° line, at 125°, (Megalith Giant/ Dolmen alignment) on its southeast direction encounters the city of Tunja. The city was built on the site of a pre -Colombian settlement dated approximately 12,000 BCE. On its 305° NW direction across the Caribbean Sea, in Nicaragua the line finds the petro glyphs of Sulutara, in Honduras the Naranjos Mounds and Cerro Palenque. In Guatemala finds the Temple and Citadels ruins of Quiriguá, Cancuen and Aguateca. The line continues onto México, and in the state of Campeche in the Yucatán peninsula it finds the Pyramids and Citadels of; Bonampak, Yaxchilán, Palenque, Villahermosa, Comalcalco and La Venta, then, across the Gulf of México it encounters the El Tajín Citadel, and the Defensa Pyramid. Further northwest, in the Baja Peninsula; it crosses over the copper and zinc mine of El Boleo. Continuing

north-west are the caves of San Francisco and San Pablo which display Paleolithic paintings. In all, the Monolith Giant and
dolmen establish a Great Arc that connects sixteen archaeological sites in its NW-SE direction and on its perpendicular NE-SW line another Great Arc connects twelve sites. An alignment of the megalith giant with a peripheral megalith at 167.42° sets the direction of a great circle which reaches the Ponce megalith in the Kalasasaya Temple in Tihuanaco and continuing south it finds the Chulpas of Curahuara and cuts through the Sajama desert where it aligns with Nodes 31, 30, 107, 38 and eighteen kiva-like structures. On its northwest direction the great circle aligns past the Newark Ohio circle and in Canada crosses by the Churchil Inukshuk. Another alignment from the central megalith to one on the northeast at an angle of 44.84° reaches the circles at Junapani, India; there a great circle in that direction aligns with circles alignment 13, see graphic on page 249.

Unlike the Great Arc Caracol-Göbekli discussed earlier, these Great Arcs do not appear to have an astronomical alignment explanation beyond that of the megalith giant with a menhir at 336.44° which is the circle of illumination at this latitude. In the graphic, p.107, the diagonal line pointing northwest shows the Circle of Illumination.

A line drawn connecting the The Lost City of Tirona with the El Infiernito, continued south connects with Nasca. The line is shown on the map on the following page -in yellow- later on we will address this line's other connections of interest. The Lost City of Tirona has its own connection with Mesoamerica. It is connected with the Temple of the Sun at Teotihuacan, in México. The white line crossing east west on the map on the next page, are actually two parallel lines with an unusual arrangement. The lines start at either side of the base of the Templo del Sol in Teotihuacan. They have a bearing of 106.15°, and are nearly perpendicular to the promenade known as the Avenue of the Dead, which starts at the Pyramid of the Moon. At the other end of the lines, in northern

Colombia, the two lines encompass the length of the steps of the Lost City of Tirona, at the foot of the Sierra Nevada de Santa Marta. This may be of significance for cultural correlation. Ms. Cielo Quintana, in her Graduate Thesis - 1979 -quotes work by

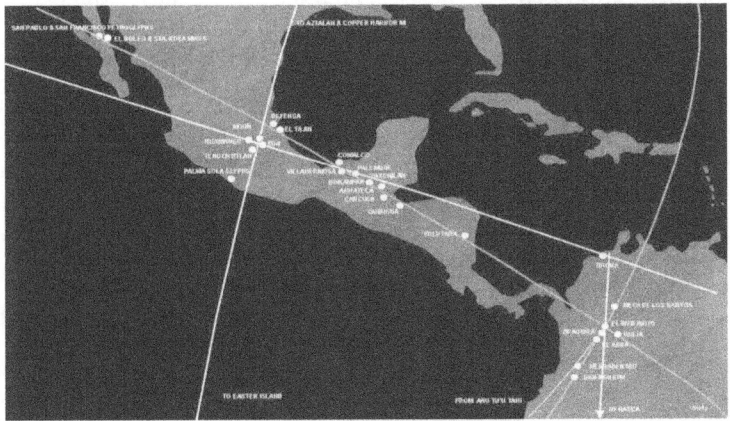

researchers; Reichel-Dolmatoff, Carlos Angulo Valdéz and Luís Duque Gómez.

"Some investigators have tried to correlate certain Tairona traits with Meso America and Central America". The Tirona Culture was renowned for its gold artifacts and the abundance of gold in the region. Fourteen century explores speak of rich mines on the north western hills of the Sierra Nevada. The Lost City of Tirona is equidistant from the Sun Pyramid and Nasca; the distance to each is 1,764 miles. Some of the other alignments between the central megalith Giant with others on the periphery are: the first one at 44.71°, in France finds the Menhir of Fohet. In northern India finds the Rankapur Jain Temple and the Junapani circles -shown on page 26 and in Indonesia encounters the Maura Takus Temple and passes fourteen miles off the Candi Padang Roco Temple ruins. In the opposite direction at 224.71° in Easter Island finds Ahu Tu'U Tahai; this is a reciprocal connection with the island. Evaluating the other megaliths clockwise; starting with a menhir at 49.34°, in Venezuela passes near the menhir alignment at Guacara.

In Turkey the line finds Göbekli Tepe. The next alignment with a megalith at 58.23° aligns with the pyramid of Chephren and the Sphinx at Giza. Following, a megalith at 62.06° in Libya finds the petroglyps of Wadi Ash-Shati and in Egypt aligns with the Temple of Luxor. Across the Red Sea near Medina finds the ruins of Wadi as Safra and cutting across southern India into SriLanka the line passes seven miles from the Sigiriya rock and over the Temples of Polonnaruwa, further southwest aligns eleven miles north of the Stupa and Temple at Budhdhangala, the next alignment with a menhir at 70.81°traces an ellipse that reaches the Pyramids of Meroé in Sudan and Jebel Bura in Yemen.

GÖBEKLI TEPE

Ahu Uru Ururanga on the east coast (27.15°S 109.34°W) connects Easter Island with Göbekli Tepe in Turkey (27.32°N -38.92°E). The Ahu has a bearing of 61.31°; a great circle with this azimuth connects the two passing through the gold region in Guyana - Cuyuni Mazaruni, one of the world's largest. Göbekli Tepe is shown below. This site is considered to be the oldest temple on earth, dated 11,500 years old, about 7,000 years older than the pyramids of Giza. The site's exploration started in 1995 by German archaeologist Prof. Klaus Schmidt. A detailed historical review of this site by Elif Batuman was published in The New Yorker magazine in December 2011. The Ahu Uru Ururanga's direction sets a great circle that crosses by the Ranu Raraku volcano the quarry source of many Moais. From this quarry site the EISP, Easter Island Statue Project, unearthed two Moais with special significance; they bear on their backs many glyphs among them two are spirals, mentioned earlier. At this point it is important to note the observation author Wayne Herschel made (on his web site), regarding the unearthed Moai:

Göbekli-Tepe. Wikimedia Commons

"The thin arms art style reaching forward with fingers pointing down at 45 degrees is exactly like the recently unearthed megaliths at Göbekli Tepe in Turkey. This alone speaks volumes and if this isn't convincing enough, the two large symbols on the back of the Moai are of the same theme and style too".

The connection between Easter Island and Göbekli Tepe is affirmed by the alignment previously described. Göbekli Tepe is one of the seventeen sites that form the Great Arc that connects the 'old world' megalithic monuments with the 'new world' pyramids and citadels as was discussed earlier in this work. There is a second set of Mayan Citadels that align with Göbekli Tepe; it will be addressed later. Göbekli Tepe, besides its historical and archaeological value stands on its own as a significant geographical reference point as will be shown.

On its trajectory from the Ahu, in Perú the circle cuts through the middle of the ancient settlements of Cañoncillo 7.402°S 79.446°W and the Huacas of Arco iris, Chan Chan and Luna y Sol 8.006°S 79.075°W; all of these within a twenty five mile radius north and

south of the circle. In Greece the circle lines up with Stratos and passes twelve miles north of Delphi and in Turkey twenty two
miles north of Hierapolis. Following the circle east of Göbekli Tepe, in India it crosses .75 mile south of the Stadhara Stupas and three miles from the Sanchi Temples.

Göbekli Tepe was the subject of a recent book by Andrew Collins; Göbekli Tepe: Genesis of the Gods, from it we reference the positioning for the various megaliths at the site. The great ellipse generated by following Ahu Uru Ururanga's bearing crosses over Göbekli Tepe atop Enclosure C - the nomenclature used by Collins. Using his and Academia.edu diagrams and an overhead photograph found in Steve Watson's Blog: (http://stevewatson.info / blog/2012/03/) for the largest of its circles - Enclosure C- we developed the graphic shown below. With this graphic, Google© earth and a drafting program we found the geometric layout of the megaliths. It was determined that megaliths 13 and 48? (blurred in the original) are aligned E-W at 90°, while 27 and 13 align at 0°N. The gap between megaliths13 and 27 was measured to form an arc of 11.4° with the E-W 90°line; that is it generates a radius at an angle of 281.4°NW; which coincides with the bearing of monolith 27. Following this line around the earth it finds Easter Island and there it aligns over Ahu Uru Ururanga, making the connection reciprocal between both sites' geometries. This reciprocity was also found in the alignment with El Infiernito. See graphic on page 111. The arrangement of megaliths 47 and 39 in Enclosure C and megaliths 19 and 32 in Enclosure D, align at an angle of 288.55°. A line in this direction corresponds with the alignment of the Megalith Giant with an exterior megalith at 49.34°. The alignment line of megaliths 22 and 38 of enclosure D at an angle of 207.61°, points directly to the Roman city of Jerash, near Amman Jordan. There, this line follows the direction of the Cardo Maximus and points to the obelisk in the center of the Forum and the temple of Zeus. Gerasa, as the Temple complex was originally known as, appears
to have been built with its axis; the Cardo Maximus, pointing in the direction of Göbekli Tepe; an unexplainable coincidence

separated by thousands of years; this is also a reciprocal alignment.

Following the north south alignment of megaliths 27 and 13, in the southern direction finds the city of Sergiopolis and the fort of Qaser al Hiyr al Sharqui and further south the petroglyphs of Al Bardh. Before reaching the Red Sea the ellipse line passes over the ruins of Wadi as Safra, crossing over Ethiopia the line passes twelve miles east of the Church of St. Mary of Zion, believed to be the site where the Arc of the Covenant is kept. This location is also found on the circle that follows the Ahu Hanga Te'e's angle. Further south in line is Lalibela 12.032°N 39.045°E with its eleven rock hewn underground churches. Aligned east following megaliths 13 and 48 is the ancient city Dara Köyü-Mardin, also Kehniya Hejîrê. Continuing the line at ninety degrees, in Afghanistan finds the AMB Temples, crossing into northern India the Baijnath temple complex and the Kandenath Gufa Shiva Temple are alingned. Following the line to Mongolia, on the border with China, is the Cave Temple Peik Chin Mayanook and further east in Thailand the temples of Nakhon Phanom also align.

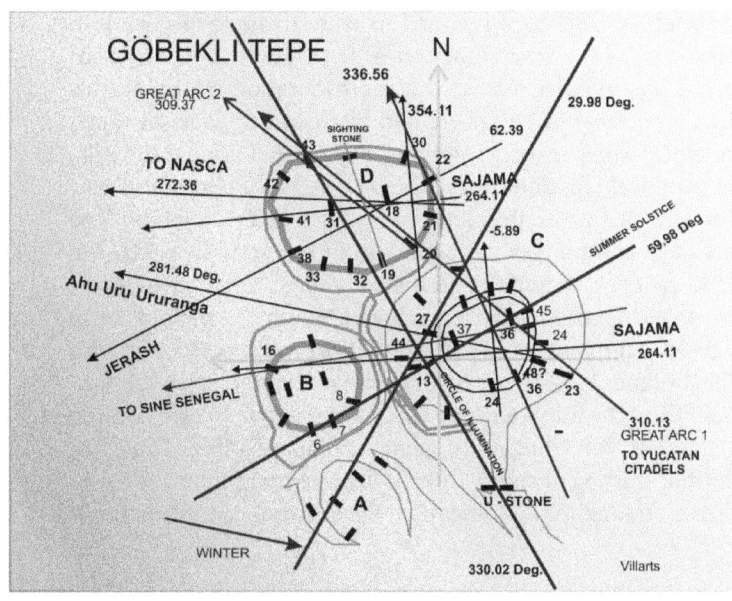

In the westward direction at 270°, in Turkey the following sites are aligned: Kargi hanAna Kapi, Aspendos, Silyon, Perge and Termessos. Continuing on to South America, south east of Cusco the Inca capital, three alignments with ancient sites are found; Panahua, the Cairns of Coropuna and the ruins of the city of Maukallakta.

There are two alignments of megaliths in the enclosures, which are particularly important. In enclosure C megaliths 24 and 28 align at an angle of -5.89° or 354.11°; the same alignment is found in enclosure D between megaliths 20 and 30. A line perpendicular to the enclosure C alignment at an angle of 264.11° aligns with megaliths 44 in C and 16 in B. This alignment when followed to South America reaches the heart of the Sajama desert where thousands of unexplained lines and structures are found. More will be said about them later. A line traced from the south side of the main megalith 18 to the north side of megalith 31 in enclosure D has an angle of 272.36°. Following this line to Nasca it finds Göbekli Tepe's line on the plain, it aligns with a parallelogram glyph and ends at a node point found at 14.693°S 75.119°W; this alignment is reciprocal as well. It was also found that megalith 36? (at about 5 o'clock location) has an azimuth angle of -23.44° in line with megalith 22 of enclosure D. This is the angle of declination of the earth's axis. The representation of the earth's axis tilt is a physical fact that we had discussed appears at many archaeological sites, except perhaps the great Pyramids. Again, this astronomical fact does not stand by it self at this site; it is part of the record of the earth's solar mechanics at this location; the structures are another geodetic marker. In the graphic, p.115, we find the *Circle of Illumination* and the summer solstice, as well.

The sunrise at the solstice has an angle of 59.98°, perpendicular to the circle of illumination which has an azimuth 330.02°at this latitude. On June 21, 2014 at 5:09am the calculated solar azimuth was 60.03° and the sun's altitude 0.03°. In the graphic we can see the alignment of megaliths that delineate the *Circle of Illumination:* starting at the U Stone it aligns with menhir 13 in enclosure C and menhirs 31 and 43 in Enclosure D. At the summer solstice the sun's azimuth aligns with menhirs, on Enclosure C: 45, 38, 37, 13. On Enclosure B menhirs: 8, 7, 6. One more interesting fact; in the graphic we can see a line which connects Göbekli Tepe with Jerash and nearly coincides with the angle of the Cardo Maximus; the angle difference is 0.45°; the extended Cardo Maximus line passes two miles west of Göbekli Tepe. These two locations share the same *Circle of Illumination* at the winter solstice, the sun angle is 61.93° at Jerash; less than two degrees difference between the two locations.

To close this analysis we look at the second Mayan Citadel alignment whose arc reaches Göbekli Tepe forming a second Great Arc. We start by describing the local geometry that sets the bearing of first Great Arc, described earlier, it starts at Abu Dahbi and through Göbekli Tepe. This line crosses through Europe; in the UK passes near Stonehenge and across the Atlantic in Yucatan, México aligns with various Maya citadels reaching the last one Tontiná in the state of Chiapas. This Great Arc's line segment through Göbekli Tepe starts at megalith 23, goes through 24 and central megalith 36 in enclosure C and connects with megaliths 20 & 43 of enclosure D at an angle of 310.13°.

For the second arc, a line that starts at megalith 20 of enclosure D goes through megalith 43 at an angle of 309.37°. Following this arc line to the Yucatan peninsula in Mexico forms the second Great Arc which aligns with five different citadels from the first alignment: Cobá 20.494°N 87.719°W, Becan 18.516°N 89.466°W, Yaxchilán 16.899°N 90.964°W, Bonampak 16.703°N 91.064°W and Lagartero 15.827°N 91.884°W. This is the fifth alignment that connects Göbekli Tepe with key archaeological sites in the Americas.

The alignments that were built into the layout of the stoneworks at Göbekli Tepe, as far as Ahu Uru Uranga in Easter Island and many other archaeological sites in between, confirm this site is one of the 'Navels of the World'.

Ahu Akahanga's bearing of 75.77°NE sets the direction of a great circle that connects the 30.4°N and 30.4°S parallels; this azimuth is confirmed by the alignment of two moais: Te Moai south of Akahanga and the Ahu One Makihi north of it. The line on its north eastward direction crosses into Perú by the town of Puerto Caballas just west of Nasca. In Nasca it passes 0.8 mile south of the Astronaut glyph and over the Pelican glyph, further inland the line cuts between Machu Picchu and Sacsayhuaman, over the Moray Inca (see photo on page 17) and the Inca Salinas of Maras.

The azimuth of the circle described here is geometrically set by Ahu Akahanga and at this angle the ellipse aligns precisely over the Pyramid of Cheops (29.975° 31.140°E) and in Iraq finds 'modern' Basrah 636BCE; formerly known as Sumer dating back to between 4500-4000BC. This alignment is particularly significant; Giza and the pyramids' alignment are discussed later in Chapter 10 dedicated to it. Between Giza and Basrah the circle passes five miles south of the Lost City of Petra (30.32°N35.44°E) dated 312BCE. It is believed one of the Petra tombs carved on the side of the mountain is that of Moses' brother Aaron. Petra is, also, aligned with the Temple on the Mount's western wall. Further east in Iran the line crosses between Apadana-Persepolis (29.93°N 52.89°E) and the Tomb of Cyrus (30.19°N 53.17°E) in Pasargadae. Further east, in Iran, the line passes near Persepolis and Naqshe Rostam 29.99°N 52.87°E and in India finds the Temples of Kuajurajo, Shiv linga and the Ravan Mada Caves. In Thailand the circle aligns the temples of: Prasat Ban Ku Moe, Prasat Khumdin and Wat Nong Ta Kian.
This alignment improves on the alignment described by Jim Alison based on his axis point distance theory[12]. This circle aligns sixteen archaeological sites around the globe. The number of alignments dependent in the Ahu's azimuth reiterates the concept of global planning for the location of all these sites.

Ahu Ko Te Riku. Some of the sites the Ahus and or their Moais lines point to are also found with Nasca lines, and the

Candelabra. Comparing the data from the three locations we find that, in some instances, their lines point to the same site or mineral resource: gold, copper, iron, salt and nitrate.

In some instances, the locations and ore types are the same. This means there is global triangulation for those site's locations. The copper belt in the Congo Zambia region (-11.78°S -27.78°E) is a prime example: the region is mapped from these three locations which are thousands of miles apart from each other.

Triangulation of a location ensures its geographical position is accurately determined. In this study we found several instances of this trigonometric mapping technique. In this instance the three lines are: from the Candelabra, a line at about 100°following the top line of the base, from Nasca the Colibrí's (humming bird) wing span azimuth 284.83°, and from Moai Ko Te Riku's line of sight at 130°. This line is shown on the right of the graphic on p.99. Now we know what Moai Ko Te Riku is looking at!

Ahu Ko Te Riku's line of sight, pointing to Congo at about 130° is perpendicular to the direction of the Ahu. The Ahu's base is trapezoidal; the ocean side runs at about 39.15° whereas the inland side runs at about 44.26°. The second ahu layer has an angle of 40.8° it is perpendicular to the Moai's line of sight. At this angle a great circle crosses over the Avebury Henge at the megalith giant. The upper platform's angles are about 34.22° and 37.43°.

A line drawn at 37.43°passes near an important archaeological site in Costa Rica: The Guayabo site (9.97°N 83.69°W); this connection can be of historical significance. The phonetic sound of the Moai's name and the number of syllables is identical to the country's name in Spanish with the local accent: Ko-Te-Ri-ku = Co(s)-ta-Ri-ca. It is tempting to speculate that when Christopher Columbus arrived in Costa Rica he may have named it after hearing the name from the natives. The natives of the region of Turrialba Valley where Guayabo is located date back 10,000 - 7000BCE.

"Accounts differ as to whether the name *La Costa Rica* (Spanish for "rich coast") was first applied by Christopher Columbus, who sailed to the eastern shores of Costa Rica during his final voyage in 1502". The Costa Rican Embassy[21].

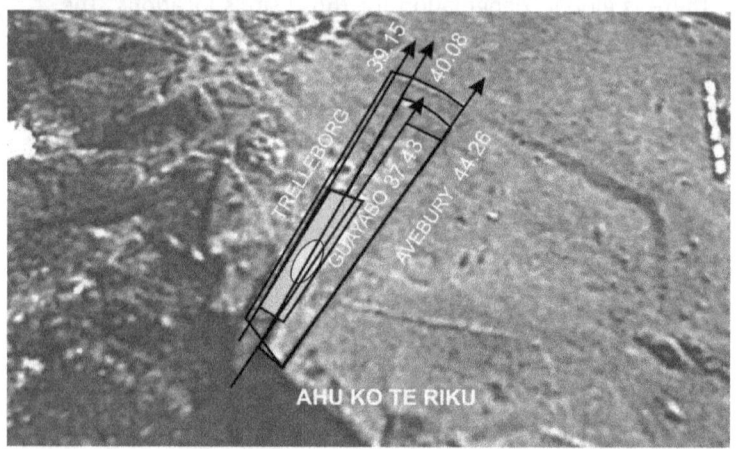

The ocean side line with an azimuth of 39.15° traces a great circle which, in Denmark, crosses 2500 feet off the Trelleborg fort and 1,400 feet off the Elverhøj mound. Another line with an azimuth of 39.13° traces a circle that aligns with the Breddysse dolmen, the Hagbard mound, the Eskildsrup dolmen and the Jættestue Maglehøj mound.

On the island there are two Ahu sites with geographical alignments: Ahu Hanga Poukura (27.17°S 109.38°W) and Ahu Vinapu (27.18°S 109 41°W). In Easter Island, as with Göbekli Tepe, the cardinal points and the earth axis' declination angle are embedded in their designs; here is built in the direction of the Ahus.

Ahu Hanga Poukura has a bearing of 66.56°. This bearing, we recall, is the angle of the ecliptic Equator, same as we have found at other sites. A line drawn perpendicular to it has an azimuth of 336.56°; equal to the earth's axis declination angle. The axis line crosses the island on its northward direction. Before leaving the island it passes near a Manavai. This structure, we recall, its cross section is a Lemniscates curve. Following the Ahu's 66.56° bearing to the northeast, we find the location of Chavín de Huántar the oldest civilization in Perú dated about 5,000 years, discovered by Julio Cesar Tello "the father of Peruvian Archaeology". Following the line further northeast, in India, finds the temples of Bijolia, Shirivi Narayan and Lingaraj.

Archaeological structure alignments with the Cardinal points are common around the world. Therefore an off-north alignment can be presumed was set intentionally and in many instances it can be confirmed as we have shown. Ahu Vinapu illustrates this clearly.

Ahu Vinapu, is located just south of Ahu Poukura. At Vinapu there are two Ahus; the southernmost one is known as the "Ceremonial center", its back wall has a 0° north azimuth; whereas its sea side wall has a 1.2° bearing. This small difference in alignment angle of the sides on the same object, indicate a purposeful design element.

Other monuments or glyphs where small alignment difference vs. north is significant are: the Candelabra, it may be recalled; this was shown via the difference in orientation between the main stem 175°and the left fork's orientation of 180°. Others are; The Temple on the Mount in Jerusalem and the Temple of Jupiter at Baalbek. The dome of the Temple of the Mount is aligned 90°with the Mount of Olives and the Pater Noster Temples. The Temple of Jupiter at Baalbeck has an east-west alignment tilted 15° to 255°. Recall, Jupiter's Temple is aligned with The Temple on the Mount through the Golan circle. The 15° tilt is significant; a line drawn perpendicular to the 225 ° angle, in a southerly direction finds Mecca. The eastern side of the Temple of the Mount has a bearing of 168.48°. Following this direction south it aligns with the historical city of Masada (31.31°N 35.35°E), while the western wall with a bearing of 173.6°, aligns with the lost city of Petra in Jordan (30.32°N 35.44°E).

<u>Ahu Vinapu-Ceremonial.</u> This is the only Ahu with 0° alignment. The only other structure in the island with this alignment is the round stones arrangement known as Te Pito Te Henua- "The Navel of The World", shown below.

At this location the difference in the headings of the Ceremonial Ahu's and the angle of Ahu Vinapu, a few steps north of it, at an azimuth of 26.57°, indicates these are important directions to follow, which we discuss in turn. Ahu Vinapu Ceremonial's circle with an azimuth of 0°, in México aligns with the Sonora Stones and Onavas; a town where elongated skulls were recently found. In the US the circle goes through the Hovenweep Pueblos in Utah.

Continuing north, the line passes near the Madicine Wheel in WY and in Canada aligns with the Cluff Lake Uranium mine in Saskatchewan (58.37°N 109.5W). The Hovenweep Pueblos are discussed in Chapter 11. Following the great circle on its southerly direction at 180° in Pakistan encounters the Temple of Jain, Nagar Tar Sindh. A line perpendicular to the Ahu at 90°, in India finds the Vaishali Temple. The eastern side of the ahu with a bearing of 1.2°reaches the Peñasco Blanco Pueblo in Chaco canyon.

Ahu Vinapu, is located northeast of Vinapu ceremonial, it has a bearing of 26.57°. This is the angle of the *Circle of illumination* at the winter solstice in the southern hemisphere at this location: the circle on this direction crosses along the east coast of the Yucatán peninsula. In Belize finds the Caracol (16.79°N 89.11°W) and the Lamanai 'Mask' Temple (17.62°N 88.7W°), shown below. In México it lines up with the Castillo at Tulum (20.20°N 87.43°W). This pyramid's location is triangulated from Easter Island, Göbeckli Tepe and Machu Picchu. A circle perpendicular to the Ahu on its easterly direction, in Brazil finds the Petro Glyphs of Sao Desiderio mentioned in Chaper 1. In Qatar finds the Al Jassasiya petro glyphs and in India aligns with the ancient ceremonial site and Temples of Beteshwar to the God Shiva. In China finds the Meng Huan Pagoda. The corresponding circle for the summer solstice (334.44°) crosses over Ahu Vai Uri on the western coast.

Ahu Akivi, the inland Ahu, has a bearing of 356° or -4°N. A line drawn with this bearing, in the Baja Peninsula, México, finds the geo glyphs of San Pablo. Inland, in Arizona it encounters the

Lamanai Temple of the Masks, Yucatán, Belize - Wikimedia Commons- Tibor Marcinek

Harquahala and Golden Eagle gold mines (33.68° 113.62°). Further north it goes over the Grand Canyon believed to have been inhabited for over 12,000 years, north of it, in Nevada, finds the Bonneville Salt Flats. A great circle drawn perpendicular to the Ahu at 86.0°, 12,200 miles away finds Moenjo Daro the early civilization in the Indus Valley dated around 2200BC. Moenjo Daro is 150 miles from the antipode point for Easter Island in the Indus Valley. On its way to Moenjo Daro, near Arica Chile the line finds the Circle geoglyph of Camarones and Acuya, it crosses 40 miles south of the ruins of Pukara San Lorenzo and Tambo Zapahuira both dated eight century BC and through nodes 87 and 88 in the Sajama region. In Egypt it finds the Kom Ombo and Abu Simbel Temples and in Saudi Arabia the ruins of Khaybar.

The Ahu Akivi Moais face the Pacific Ocean at an azimuth of 266.0°. The calculated sunset for September 23, 2015 -the spring equinox- will be at 1716hrs, the sun's azimuth will be 264.53°, the moais will not be facing the sunset precisely as it is sometimes claimed; they will be 1.47° off. The four degrees off north the Ahu is aligned to make the Moais face Moenjo Daro and the other sites mentioned above, probably, not the sunset.

Ahu Tongariki. The Ahu has a north-east heading of 30.45°, the angle for the Circle of illumination at this location is 25.56°; in theory at the culmination of the summer solstice the sun aligns perpendicular to the Circle of illumination at an angle of 115.56°. At sunrise on December 22 at 6:20 am the solar azimuth was 109.35°. The Ahu does not receive sunshine perpendicular to its heading at any time; sunrise or sunset.

Ahu Tongariki's Circle of illumination is shared with the east coast of the Yucatan peninsula, México; all the archaeological sites, temples and pyramids on the coast are illuminated at the same time as are all archaeological sites along the US eastern seaboard. In Florida are the Crystal River Mounds, in Georgia the Sapelo shell Ring, in South Carolina the Spanish Mount, in Vermont Chimney point, in New Hampshire America's Stonehenge and in Maine the Solon Petroglyphs, in Canada the recently discovered archaeological site near Rigolet. On the same Circle of Illumination, twelve hours earlier, in India and Sri Lanka a group of temples receive sunlight at the same solar time. In India: Persnavath Mandir, Ranchhodrayji, Saptashringi Devi, the temple Complex in the Anegundi region, discussed earlier regarding Machu Picchu, Lepakshi, Kanjamalai Perumal and Sriranganathaswami and in Sri Lanka, Sankapala.

A line perpendicular to the Ahu's direction of 30.45° passes one hundred miles west of the Ta'Er Lamasery (36.49°N 101.56°E) in Xining China, whose history dates back to around 200BC. This Buddhist complex is of historical significance; it is the depository of Buddha's writings. Further north it encounters the Quaidam Basin and the Helan Mountain petroglyphs and in Beijing finds the Heaven Temple.

Following the Ahu's line in the southerly direction 210.45° reaches the eastern coast of Sri Lanka. Nineteen miles from the coast is the Budhdhangala Stupa and north of it forty miles from the coast is the ancient Sacred City of Polonnaruwa with its main Stupa Rankoth Wehera.

The Stupa aligns with the Bodhigara stupa, the Samahal Prasada Temple and the Gal Viharaya Temple on the hill at an angle of 7.88°; the Polonnaruwa complex runs for 1.6 miles in this direction. East of Polonnaruwa is the Lion Rock with the Sirigiya Fort on top. The moat around the fort and citadel is aligned at 7.88° as well. The pathway that connects Sirigiya with the statue of Buddha in Ehalagala a mile west of it runs perpendicular to the moat at 277.88°. Following this direction it finds the Aukana Buddha, a monolith carved on the side of the mountain. In Ethiopia the line passes eleven miles south of Laibela and in Nigeria near the 'Holes in Granite' rock on Mt. St Ives.

The Polonnaruwa alignment has a global scope. Following the ellipse at the 7.88° angle, it aligns with sites in India, China, Russia and North America: India- the temples of Vigneswara, Bhimeswara, Dubramanyeswara and Shiviri Narayan.

In China the menhir and cairns at Heavenly Lake, in Russia the circle passes by the Salbyksky Mound 53.894° N 90.772°W. In

North America it finds the Pictograph Cave in Montana, the Medicine Wheel in Wyoming, and the New Mexico Pueblos of; Poshouinge, Puye, Tsankawi, which line up precisely and the Alcove and Tyuonyi pueblos. Crossing into Mexico it finds the Petroglyphs at Cerro Blanco. On its southerly direction in south Sri Lanka the Temple of Sankapala is also in line. Polonnaruwa aligns with the Mehgalith Giant at El Infiernito, as described earlier.

Ahu Vai Tara Kai on the north shore has an east-west bearing of 94.65°. On its westward direction it reaches the island of Guam where rows of megaliths are found, which are strikingly similar in shape to Moais with their Pukaus on, known as Latte Stones; however they do not have human features. These megaliths range in height from about three feet to the largest of about ten feet; shown below.

Ahu Atanga, is also located on the north shore, with a heading of 260.23°. A great circle drawn in this direction connects Easter Island with the Island of Tonga where the Ha'amonga'a Maui Trilithon is found; shown above. These two sites and Ahu Akivi's alignment with the Island of Rarotonga are connections we found between Easter Island and the Pacific Islands. Archaeologists theorize the peoples of Rapa Nui are of Polynesian ancestry.

The Baja Point Ahu has a bearing of 81.78°NE. A circle with this bearing, in Southwest France finds the Carnac region, mentioned earlier, where numerous menhir alignments are found. At this angle, a line in this direction coincides with the azimuth of one of the alignments at Kerlescan (47.60°N 3.05°W), shown below.

Kerlescan Alignment-Wikimedia Commons

Ahu Tautira in Hanga Roa beach, has a bearing of 23.78°, a line drawn from it following this direction, in the Yucatán peninsula, connects four Mayan Culture Citadel sites: Lagartero (15.827°N 91.88°W), Chincultik (16.13°N 91.79°W), El Tigre Itzankanak (18.121°N 90.838°W), Ezná (19.61°N 90.24°W), Kanki K'nish (20.0°N 90.11°W), Xcalimkin (20.17°N 90.01°W) and Uxmal (20.36°N 89.77°W). Across the Gulf of México in Georgia, US the ellipse finds the Ocmulgee (p. 129) and the Rock Eagle Mounds.

Ahu Te Peu's line with a bearing of 15°, in Veracruz México, finds the archaeological sites of Yohualichan (20.06°N 97.5°W) and El Tajín (20.45°N 97.38°W). El Tajín, we recall, is also found on the El Infirnito Great Circle 305° Line. In Kansas, US it finds the Graham Cave where human s lived 10k years ago. In Wisconsin, west of Aztalan crosses by the Man Mound and in Canada passes fifty miles from the Hudson Bay Inukshuk.
In India passes eleven miles from the Ayodhyapuram Temple, within thirteen miles of the Damashan and Tambdi Surla temples. South of the Ahu the circle aligns with the O Pare Rock near Tahai.

Ocmulgee Mound Photo by the Author

Ahu Hanga Hahave with a bearing of 72.83°, points to historical locations in the Middle East. A circle at this azimuth, in Israel finds the Temple on the Mount. Following into Jordan, it finds an area akin to Kirbeth Al Umbashi, eighty nine miles southeast of it; with structures similar to the 'kites', located at 31.840°N 37.456°E.
Further east it finds the five thousand year old city of Uruk and crosses about fifty miles north of Basrah. Basrah is one of the cities on the Caracol - Göbekli Great Arc and on the Ahu Akahanga great circle over the pyramid of Cheops. Continuing, east of Persepolis the 500BC fortress of Arg-e-Rayen is found, which is also on the path of the "Silk Road". The Silk Road was the trading route that started in china in Ch'ang-Gan -Siam ending in the Mediterranean Sea at Tyre. "Extending 4,000 miles (6,437 kilometres), the Silk Road gets its name from the lucrative Chinese silk trade which was carried out along its length, and began during the Han Dynasty (206 BC – 220 AD)" Wkipedia
Ahu NauNau on the northern coast has an azimuth of 70.0°. A great circle aligned at this angle cuts through Marcahuasi (11.78°S 76.57°W) Perú, a renowned location for its unusual rock formations, claimed to be a 'portal' to other dimensions. One of the rock formations looks like a Moai. In northern Africa the circle passes over the ancient Greek cities of Ptolemais and Cyrene. Across the Mediterranean Ocean crosses south of the port of Tyre

the end of the 'Silk Road' in Lebanon. East of Tyre in Jordan finds Kirbeth Al Umbashi and in Iraq It continues onto the ancient cities of Babylon -1894BC- and Qaryat Wasit in Iraq and the citadel of Chongha Zanbil in Iran, continuing east, it passes thirty one miles north of MoenjoDaro. Continuing into northern India it finds the Keshori Patan temple and across the Bay of Bengal in Mayanmar the Shwemokhtaw pagoda. Further east in Thailand passes near Wat Arun and in Cambodia within fifteen miles of Phnom Chisor and Phenom Dar.

Ahu Tarakiu's great circle with an azimuth of (65.88°) aligns with Tinyash in Perú, in Malta it passes twenty one miles south of the Tarxien Temples, twenty seven miles south of Apolo's Temple in Rhodes, over the Afendrika Ruins in Nicosia. In Syria finds Ugarit the site of where the first 30 letter cuneiform alphabet was found in clay tablets, the basis for Arabic and probably the Latin and Greek alphabets, also in Siria it finds the ruins of the 2nd AD Roman Temple in Ithriyah (35.367° 37.779°). In India finds the temples of Bijiola, Madakini, Shivri Narayan, Lingaraj and the Sun. In South Sulawesi, Indonesia crosses seventeen miles north of the Tana Toraja Burial Site.

We have shown that the alignments of Easter Island's Ahus; their location and headings follow a design. It was also found that the angular rotation of these alignments around the island is similar to that we had found for the Nasca lines. As in Nasca, the Ahus and Moais point in all directions 360° around. The data plot, on page 133, illustrates this. We cataloged nearly fifty alignments at Easter Island and have discussed the ones we thought their orientations showed the scope of their significance in a global context.
We have found that the data adds to the previous results, which make Easter Island a building block, part of an ancient global map; a map that, in some instances, guided us to archaeological important sites which serve as hubs from which other alignments emanate.

We began this part of our research with the alignments in the Middle East. These alignments connected the continents through the Great Arc expanding from Abu-Dhabi in the Persian Gulf to the Maya city of Tontiná, about one hundred miles from the Pacific ocean. The connection to the Americas brought us to identifying the Candelabra and Nasca as places belonging to a distinct group of archaeological sites with unique design characteristics. It was discovered that the Nasca method for finding and mapping a site, is based on a systematic arrangement of lines with their azimuth headings originating at Nasca. Also found that the same method was used in Easter Island. Because of its terrain, the island, appears, had a limited capacity to accommodate all the alignments that were needed. In the Nasca plain this was accomplished by eliminating the need for Ahus and Moais which were replaced with lines.

Easter Island would seem was the original starting point for the mapping of locations around the world with the apparent purpose of preserving the record in an indestructible manner. The island's connection with the Pyramid of Cheops is of particular significance, as are the connections with the Acropolis, the Temple on the mount, Jupiter's Temple at Baalbek, Göbekli Tepe and the Golan Circle.

The surveying method; viewing the site as a radial arrangement hub was extended from Easter Island, to Nasca, to El Infiernito's Megaliths arrangement and beyond. Regardless of layout this radial alignment pattern repeats at several other archaeological sites in the Americas, Europe and Asia as will be discussed in the following chapters. The designers of these sites, regardless of who they were succeeded at leaving a long lasting record, in many forms most of them virtually indestructible!

The Easter Island site has a limited number of statues Moais with names; are these statues representative of the giants reported throughout history? Are these the giants disclosed in Lama T. Lobsang Rampa's book, The Third Eye? Were these a group of Ancient Explorers?

In the survey of the sites presented thus far, we have noted the importance and the ability the ancient architects had in designing and building monuments following geometry which reflects the earth's planetary mechanics. This was seen in the structures; both arranged to align with and to have engineered into their designs the angle of declination of the earth's axis, as the earth cycles through the solstices. These facts leave us with no doubt they had a clear understanding of astronomy and the azimuth of the Circle of Illumination at the various locations. In the next Chapter we tackle this subject and present an in-depth analysis of how this technology was used to accomplish the greatest achievement of that ancient civilization: A GLOBAL CLOCK.

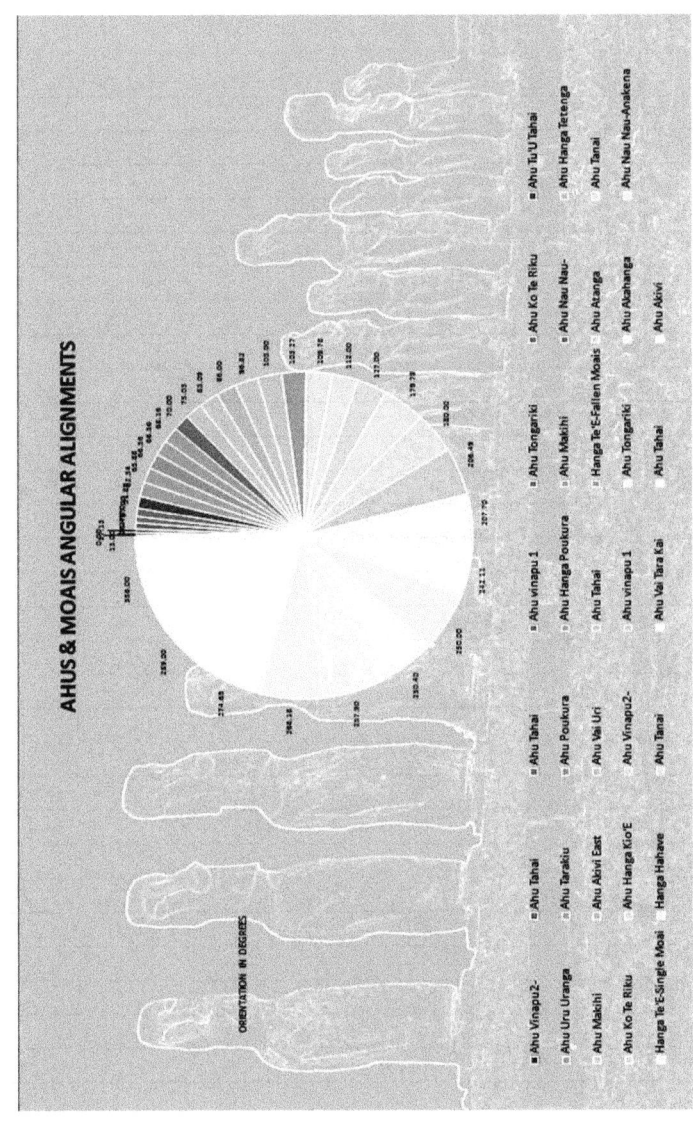

Notes

CHAPTER 6

THE ANCIENT's GLOBAL CLOCK

The way the sun illuminates ancient structures and their sites, stone structures and those with glyphs which make them particularly notable, have tantalized the imagination and curiosity of scientists and the population at large since the beginning of recorded time. This fascination results in a variety of theories that attempt to explain why some of these structures align in various ways at the solar solstices and what the purpose of their alignments might have been. As to purpose the explanations range from; fantasy to the mystical: Indiana Jones finds the location of the Arc of Covenant when the sun aligns in a particular way, mysticism is exemplified by the Druids who were "concerned with the stars and their movements, the size of the cosmos and the earth, the world of nature, and the powers of deities" as described by Julius Cesar[36]; the Druids performed their rituals at megalithic sites. Finally, technology brought astronomical measurements that gave rise to the science Archaeoastronomy which serves to confirm the structure's alignments with the stars, but still leaves the reason **'why'** to be debated.

In this chapter we present new findings from the continuation of the earlier part of the study, which had shown evidence the arquitects of many archaeological monuments, knew the earth's axis tilts 23.44° with respect to the planetary ecliptic.

The earliest recognized evidence of this knowledge is attributed to the Greek philosopher Oenopides 480 BCE[37]. The continuation of this work now takes us beyond the knowledge the Ancient Architects had of a single astronomical fact, now we know it includes the measurements and mechanics of the seasons and of time. This part of the research reveals the ancient architects had a universal vision and the technology to quantify astronomical events. They had the where withal to take that knowledge and to cast it in stone; as an ANCIENT GLOBAL CLOCK!

ORIGINAL DISCOVERY

Earlier in chapter 4 we presented the discovery that lead to finding the purpose for the famous Nasca Lines: The lines point in the direction of other archaeological sites. That is; Nasca is a cartographer's Rosa de los Vientos, the star-like compass found in many maps, in particular old ones. It is the ancient travelers' record. Recall, on the Nasca plain there are two are predominant lines of the possibly thousands of lines reported. Those were the lines we focused on at the onset of that research. The next seven pages re-capture the alignment descriptions of Sacsayhuaman and Machu Picchu as well as the Circle of illumination concept.

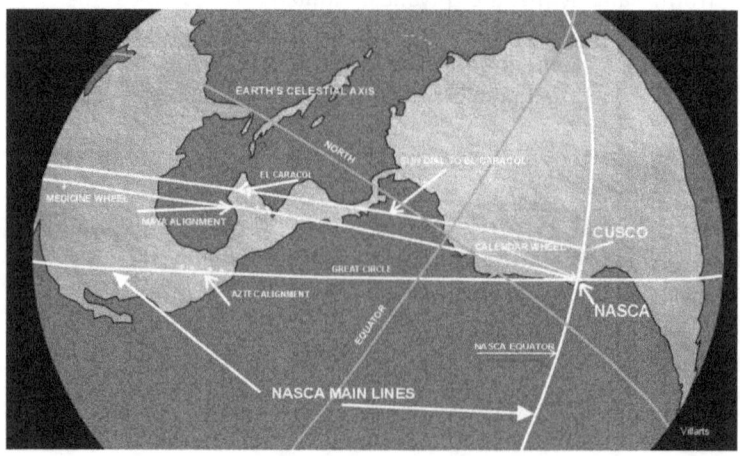

The reader may skip to page 140 and pick up the argument starting with 'Global Scope'

Measuring the main line angles we found these lines are similar in direction to the earth's axis and Equator's tilt angles at the summer solstice in the southern hemisphere. Now we revisit this earlier discovery from a new perspective. The earth's axis obliquity angle is the key factor in the changing of the seasons as the earth completes one revolution around the sun. During this time the earth's axis angle shifts from 23.44° to -23.44° (the explementary[46] angle's value is 336.56°) with respect to the ecliptic, thus the sun traverses across the Equator as the seasons shift from spring to summer and from autumn to winter; the so called vernal and autumnal equinoxes. These are the points when the angle of the sun with respect to the Equator is 0°, as shown in the graphic below.

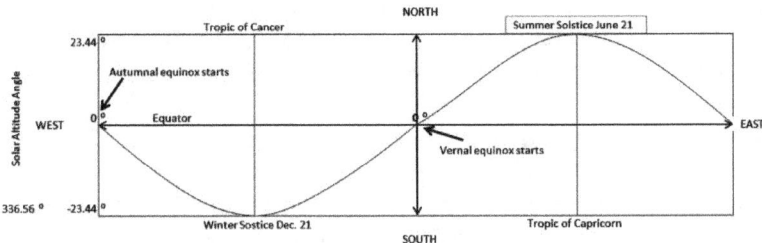

The apparent solar shift (the earth is the one shifting) causes the sunrays to illuminate the earth at a continuously changing angle as measured from any location on earth. If the sun were a laser pointer it would trace a sinuous curve onto the earth like the one shown above. However it is not, so; it illuminates half the earth at all times; second by second as it spins about its axis. The exposure to the sun is the strongest at points on earth between the Tropics of Cancer and Capricorn where the solar angle can reach maximum altitude of 90°. The leading edge of light perpendicular to the sun's center, at a given latitude, illuminates the earth's sphere from top to bottom. This line of light is known as the *Circle of Illumination,* as previously described. This line of light has a different azimuth above and below the Equator. Therefore, the sun angle is different at each latitude.

ARCHAEOLOGY AND THE CIRCLE OF ILLUMINATION
Earlier we described line on the Nasca plain with a bearing much closer to the obliquity angle of the earth's axis; it has an azimuth of 336.84°. This line is closer in value to the actual azimuth for the *Circle of Illumination* at Nasca's latitude. At this latitude its value is 335.71° at the summer solstice in the southern hemisphere; both lines are shown in the graphic below, as they appear on the Nasca plain. The main lines and others similar to them appear to be a record the sun angle(s) at Nasca's latitude; the similitude in value to the earth's axis and Equator appear not to be coincidental.

Structures or places lined up along ellipses at the angle of the *Circle of Illumination* receive sunlight at the same time, i.e. the sun rises and sets at the same solar time -assuming there are no obstructions. The Arctic and Antarctic Circles are imaginary circles formed by the set of ellipses drawn by the *Circle of Illumination* as the earth spins.

Likewise, the 'ecliptic *Equators'* -ellipses drawn perpendicular to the *Circles of Illumination* at a point equidistant from the poles generate the imaginary parallel lines known as the Tropics of Cancer and Capricorn circles. These circles represent the maximum solar altitude at the summer Solstices in both hemispheres respectively. The radial lines shown in the graphic appear to be the record of the historical shift in the earth axis' obliquity angle. In the earlier discussion we showed that several archaeological sites are in line with the Nasca main line (white-324.86° NW); shown in the first graphic above labeled "Aztec Alignment". In Perú among them are; Tambo Colorado, the Band of Holes and El Paraiso Ruins. Across the Pacific Ocean, in México are: Monte Alban, Cholula, Tenochtitlán, Teotihuacan and Tula.

This alignment was the first hint for a possible rationale that could explain why the location of each structure was selected: At each location, the structures receive sun light at approximately at the same time. They mark the *line of sunlight*, like pins on a map marking a route.

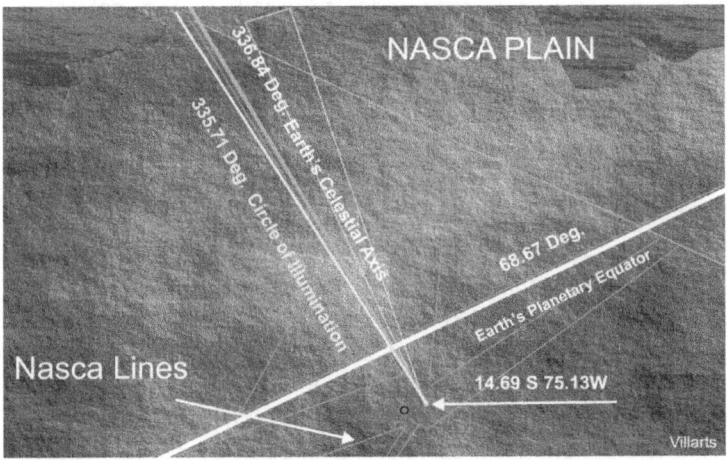

Some of these alignments could be coincidental; however, following the Nasca main line in the NW direction around the world (324.86°) to form a great circle, it closely aligns with the temple of Angkor Wat and more precisely with the temples of Pre Rup and Sien Reap in Cambodia, the antipode region to Nasca. The alignments in the antipode region also receive sunlight simultaneously twelve hours earlier; this adds support to the argument. Similar alignments on the same circle directly opposite, half a day cycle earlier, makes this arrangement unlikely to be coincidental. The angle of the *Circle of Illumination* at Nasca's latitude is 335.71°; a new circle was traced at this angle around the world. This circle is also aligned with monuments in Central America, North America and in Cambodia: The sites of Uaxactun in Guatemala, The Caracol in Belize, the Calakmul pyramid in México, the Medicine Wheel in Wyoming US, and in Alberta Canada the Writing-on-Stone and the Whakpa Chug'n Buffalo Jump sites -labelled "Maya Alignment". In Cambodia the Temples of Boeng Mealea, located forty miles east of Angkor Wat, Prasat

Kho Ker, Prasat Khao Phra and the Holy Place at Wat Phu Lon; these alignments parallel the layout of the previous set.

Following the other Nasca main line at an angle of 68.67°, it crosses atop the Sacsayhuaman Citadel in Cusco Perú, the seat of the Inca Empire. In Sacsayhuaman, known as the Inca 'Navel of the World', atop the citadel there is a circular structure in stone similar to the Nasca Calendar Wheel and the Medicine Wheel in Wyoming, known locally as "The Cusco Sun Dial", i.e. an observatory.

The Sun Dial has double row stone radii; one of the radii runs NW at an angle of 334.29°, close to the *Circle of Illumination's angle* and another nearly perpendicular to it. The angle of the *Circle of Illumination* at this latitude is 335.81°. The line which runs at 334.29° NW connects 'The Sun Dial' with the El Caracol Observatory in Chichén Itza in the Yucatán Peninsula, México; parallel to the "Maya Alignment" line. The 'Sun dial' is a geodetic marker for the astronomical *Circle of Illumination*. Sacsayhuaman is a site with a monument that encodes astronomical measurements in its geometry and placement: its location is found east on the 68.67° NE line that crosses the Nasca plain, its Sun Dial structure encodes the angle of declination of the earth's axis and a line traced at that angle connects with the El Observatorio at Chichén Itza in Yucatan, México. Similarly, the El Caracol Observatorio is aligned with the Castillo, also in Chichén Itza, following their *Circle of Illumination* at an angle of 25.17° at that latitude. This presents another piece of evidence which reinforces the proposed argument, as to why these sites were selected to place geodetic markers in. The alignments described and the displays of structural features that point in the direction of the angle of declination of the earth's axis, begin to solidify the idea that the ancient architects had a plan; one that involves Nasca and Sacsayhuaman and as pointed out earlier, also includes other archaeological sites around the world. At several of these locations the theoretical *Circle of Illumination* angle and the actual value for the angle at the location's latitude are both embedded in their designs. To support this rationale further and to delve into the true value of this discovery we need to look at Machu Picchu, where we found the entire citadel was built around these concepts.

MACHU PICCHU

Machu Picchu is located 1.5 miles high in the Andes, northwest of Sacsayhuaman at about 302°. The citadel sits on a mountain ridge that runs northwest at 337.83°, coincidentally the ridge's direction almost equals the angle of declination of the earth's axis. The Intihuatana Stone 'The Sun's hitching Post' is a structure atop the citadel that aligns with the sun at the summer solstice - another Sun Dial. The stone lines up at 336.56° with another structure atop the ridge. Also, the buildings' walls across the plaza and its terraces were built at the same angle. The *Circle of Illumination's* angle which at this latitude has a value of 335.89° is found in some of these stepped terraces. This reinforces the idea, the ancient architects were aware of how the sun illuminates the earth and the shift in angle through the equinoxes.

The foregoing indicates that the site for Machu Picchu was likely selected, because the mountain ridge follows the approximate angle of obliquity of the earth's axis. The city architects planed the structures to correct the azimuth to more closely follow the declination of the earth's axis and also included the exact *Circle of Illumination's* azimuth in the stepped terrace's design. The Sun Hitching Post stone's location (13.162°S 72.545°W) aligns with the sun at its zenith when it reaches the solstice culminations, at the spring and autumnal equinoxes; in the southern hemisphere on September 23rd and March21st.

The value of the angle for the *Circle of Illumination* at Machu Picchu shows a difference of 0.67° vs. the declination angle of the earth's axis; the difference is due to its location thirteen degrees latitude south. Because of this small angular difference the Intihuatana Stone or the Sun's Hitching Post casts a shadow at *clock time* noon at the summer solstice, however small it is; the sun angle is 4.38° degrees at noon. At *solar time* noon the sun angle drops to 0.0°, thus casts no shadow, the sun's altitude is 77.07° [41]. These metrics confirm the architects of this site had specific knowledge of planetary mechanics, as will be reiterated later. In the graphic above the lines cross at the Intihuatana Stone. The terraces and the various structures follow the direction of the white line pointing northwest; they are aligned at 336.56° and at 335.89°. The map above shows the general direction the natural ridge follows, a close up is shown on page 80. An ellipse drawn at this angle in the Yucatan peninsula finds the temples of El Castillo near Tulum and Nonoch Mul in Cobá. The same ellipse in its southern direction crosses near the ruins of Chilenoyoc and further south Pachamarca. Near the city of Arica it finds the Pukara San Lorenzo, south from there, in Chiza finds an archaeological site with anthropomorphic glyphs, continuing fifty miles south finds the famous Atacama Giant glyph and crossing into Argentina encounters the Pirichao archaeological region. On its way to Indochina in Indonesia the ellipse crosses forty eight kilometers from the ancient temple ruins of Padang Roco.

Continuing the line, in Viet Nam it finds the Chiên Đàn trio Champa Temple. All these sites are on the same *Circle of Illumination*.

A circle perpendicular to the *Circle of Illumination* at an angle of 65.89°NE -the sun angle at sunrise during the summer solstice- starting at the Intihuatana stone, on its north easterly direction in northern Africa on the shores of the Nile in Egypt, crosses fifty kilometers north of the Temples of Seti. In India passes seventeen kilometers from the Stupas of Stadhara and Sanchi 23.4788°N 77.739°W, further east it finds the Nakta Mandir Temple. In Myanmar the line crosses over the Standing Golden Buddha atop the Ma Dey Mountain and twelve kilometers south of The Golden Rock (17.483°N 97.098°W). In Thailand it finds the Phimai historical park and the Prasat Phanom Wan Temple, on the border with Cambodia finds the temples of Prasat Khao Phra, or Preah Vihear and Prasat Neak Buos. The sun follows a sequential path marked by all these structures at the summer solstice at each location, the solar angle reaches a maximum of 89.96° at Stadhara-Sanchi, then it declines to 39.74° in winter at Machu Picchu.

Machu Picchu reaffirms the concept of *purposeful design;* its layout embeds the relationship between the circle of illumination angle and the earth's axis tilt. Machu Picchu's extraordinary arrangement gives the site unparalleled importance, not only for all its archaeological attributes that have fascinated humanity for centuries, but in particular for its *tie* with the Sun-Earth mechanics explicitly embedded in the Intihuatana Stone and the architecture of its buildings. The other two sites provide additional points of reference. The illumination data for the three sites reveal a pattern previously unknown.

GLOBAL SCOPE

Nasca and Machu Picchu display the *correct* value of the angle of the earth's axis tilt at the Equator, although it has a different value at each of these locations' latitude s. However the *correct* angle value is also displayed for each site's latitude, thus highlighting the difference. Sacsayhuaman provides additional perspective; a line which connectsthe Sun Dial with the El Caracol Observatory has an azimuth close to the angle of *Circle of Illumination's* azimuth at that latitude. The connection with the El Caracol is the second connection with the northern hemisphere, the first one is between the Calendar Wheel in Nasca with the medicine Wheel in Wyoming, US; shown in the graphic on page 136. These other two 'observatories' connect with a line at an azimuth of 335.27°. All these connections broaden the focus from the regional Inca structures to include the Maya and Aztec structures and the American Indian Pueblos. The Pueblos' architectural designs and relationship with Nasca are discussed at length in chapter 11.

The *Circle of Illumination* angle increases towards the Equator at latitudes above and below until it reaches the maximum value of 336.56° at the Equator. Archaeological sites at latitudes between Machu Picchu and the Equator were evaluated; those found at short distances north of Machu Picchu had small angle differences from those previously mentioned; the angles measured increases with latitude until reaching the site of La Maná. The Maná region is the archaeological location closest to the Equator approaching it from the south; it is the archaeological region where the Eye Pyramid was found (a small stone pyramid with an obsidian eye incrusted in it that glows under black light. The pyramid is virtually identical to the picture found on a dollar bill; La Maná is located at 0.942116°S 79.213070°W. At this site the Circle of Illumination angle value is 336.55°. This location is north of the Humming Bird Pyramid (0.9978°S 79.2283°W) discovered on November 17, 2013 by Alexander Putney and Susanne Benoit near the Calope River in Ecuador[38]. This is the same immediate region where the Cueva de Los Tayos[37] is located (1.933°S 77.792°W), where many other artifacts were found. Following is an analysis of sites north of the Equator.

At the archaeological site of El Infiernito in Colombia, discussed earlier - latitude 5.647°N, one of the radii drawn between the central phallic megalith giant and a menhir on the periphery has an angle of 335.90°and the alignment of two other menhirs set an angle of 23.34°, both values are close to the angle of declination of the earth's axis. At this latitude the *Circle of Illumination* has an angle of 336.44° and 23.56° at the point where the sun falls perpendicular to it at the summer and winter solstices respectively. We found other archaeological sites in Colombia at latitudes between the Equator 0° and El Infiernito 5.647°N, namely: The Paleolithic burial caves of El Abra (4.59°N) and Segovia (2.574°N), and the Archaeological Park of San Agustin (1.915°N). San Agustin is "a UNESCO site which extends over an area of over one hundred hectares and comprises 600 known megalithic statues and forty burial mounds dated between 1-900AD"[40]. San Agustin is the closest archaeological site to the Equator approaching it from the north; its *Circle of Illumination* angles are 336.54° and 23.46° at that latitude.

The *Circle of Illumination* shift in angle between the summer and winter solstices, cycles through a point where the sun appears in line with the Equator at the Vernal and Autumnal equinoxes. Based on the data collected it was logical to expect that an Archeological site with a *Circle of Illumination* angle equal to the maximum declination angle of 336.56° should exist, its location corresponding to the inflexion point of the equinoxes; it would be located exactly at the Equator. We found such a site, it has many menhirs some of them with glyphs, near the town of Aur Duri in Indonesia, located at 0.038167°N 100.480350°W, about 4.4 km (2.7 miles) north of the Equator. The table below shows the latitude and the solar *Circle of Illumination* angle for the various sites just mentioned.

The location of the Aur Duri menhirs close to the Equator[62] marks the point of inflexion of the equinoxes; this is an important fact that speaks of deliberate planning. Another important fact related to the planning for this location is that the menhir's antipode point is located at about one hundred kilometers northwest of the La Maná region in Equador. Although this region is located twelve hours away, as expected, the Circle of Illumination's angle for this site falls in place in the data sequence shown below. There is one more fact that may connect the menhirs in Indonesia with the pyramid in Ecuador. Archaeologist Alexander Putney who

The angle values shown are for the summer solstice in both hemispheres

Site selection		Illumination Degrees	latitude°
Nasca	Line Node Point	335.71	-14.697
Sacsayhuaman	Sun Dial	335.81	-13.510
Machu Picchu	Intihuatana stone	335.89	-13.160
La Maná	Eye Pyramid	335.55	-0.942
Aur Duri	Menhirs	336.56/23.44	0.038
San Agustin	Menhirs & Dolmen	23.46	1.915
Segovia cave	Burial site Glyphs	23.47	2.574
El abra cave	Burial site Glyphs	23.51	4.590
Infiernito	Ctr. Giant Megalith	23.56	5.647

discovered the pyramid makes an important observation in a video[38], regarding the composition of the stones that cover the pyramid; paraphrasing "The composition of the stones that cover the face of the Hummingbird pyramid is similar to that found at excavations at archaeological sites in Indonesia".

Furthermore, the *Circle of Illumination* ellipse traced from the Aur Duri menhirs around the globe at an angle of 23.44° crosses near the Hummingbird pyramid, over La Cueva de los Tayos and the site where the Eye Pyramid was found. The El Infiernito shares the circle of Illumination with Gunug Padang a Temple of standing megaliths in Indonesia about 700km east of Aur Duri; more on the significance of this will be explained later.

STONEHENGE

Among the structures which encode the *Circles of Illumination* in their designs, one outstanding example is Stonehenge. In the graphic below, these circles are shown as crossing lines: the *Circle of Illumination* with an angle of 320.61° at this latitude, aligns with megaliths 9a, 9b, 80 and passes between megaliths 22 and 23 -using Anthony Johnson's nomenclature[39]. It is the white line marked with a sun, in the northwest quadrant.

The white line perpendicular to it has an azimuth of 50.61°, the same angle of a line connecting the Heel Stone and menhir 80, the Altar Stone which, as is well known, is the line of sight of the

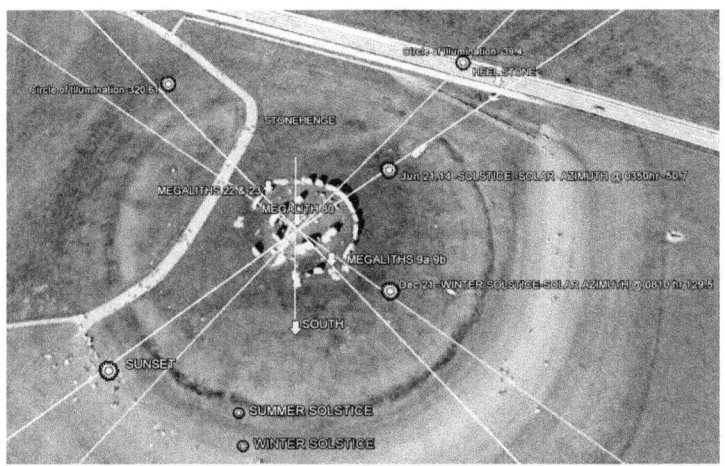

Stonehenge solar angles: Graphic by the Author over a Google© earth map

sun at sunrise at the summer solstice. The sunrise on June 21, 2014 was at *Solar Time* 0350 hours; the sun's altitude angle was 0.05°and the azimuth 50.7°.

The line at an angle of 39.4° is the winter solstice *Circle of Illumination.* The line perpendicular to it is the sun's direction at sunrise on Dec. 21, 2014 at *Solar Time* 0811 hours; the sun's altitude angle was 0.07° and the azimuth 129.5°[41].

Sharing Stonehenge's Circle of illumination at the summer solstice, in the UK, are: the Lugbury menhirs and the Clynnog Dolmen, in Ireland, the Dunnamore Dolmens, the Beaghmore Stone Circles and the Gortnavern Dolmen and Standing Menhir. Southeast of Stonehenge in the island of Sardinia, eight Dolmens align with the *Circle of Illumination*: Grotta, Chiaramonti, San Nicola, Sa Conca e s'abba, Orotelli, Mazzozzo, Mamoiada, and Tomb of the Giant in Madau.

Stonehenge may be viewed as a sextant which points to diverse archaeological places in several directions and also to the stars; among the diverse monuments it points to are; several in the Inca region in Perú and in Japan the Yonaguni pyramid. See chapter 9 The angle of the circle of illumination at Stonehenge is encoded in the same manner as it is in Nasca, Sacsayhuaman's Sun Dial and Machu Picchu, i.e. in the geometry of its structure.

ASTRONOMY

Machu Picchu's topography and architecture pointedly show the importance the ancient architects gave the *Circle of Illumination*; this is reiterated repeatedly around the world, as discussed. The relationship between the sun and archaeological structures has been celebrated, pondered about, mystified and in many instances misunderstood. The above analysis provides new insight to the logic behind archaeological structures' locations and exposure to the sun. The *Circle of Illumination* adds significance to each location. Based on it, the interest that traditionally is given to these structures regarding the way they are illuminated at the culmination of the solar solstice is reinforced and justified. However, this research shows the basic underlying reason for the alignment.

The ancient architects encoded in some of their designs the actual metrics for the *Circle of Illumination* at each location's latitude; sometimes on a single structure or in the alignment of several structures. The set of all the structures' *Circles of Illumination* from around the world, when drawn on the earth reveal a global pattern. Recall that the Aur Duri menhirs site is exactly antipode to the region where the Inca and other civilizations or their ancestors built the various sites shown on the table above.

The relationship between these two locations, we found to be the key regarding the answer to the basic question: **why** were the structures built at the locations they are found in. The alignment of structures with the *Circle of Illumination* provides coherent daytime hours at the various locations since they receive sun light simultaneously. This effect in the western side of the Americas is echoed twelve hours earlier on the same ellipse in the Far East at its antipode point; not once but twice, as explained earlier.

Those examples and the AurDuri-Maná arrangement show similar phenomena echoed on a nearly twelve hour cycle. Are these unique? We set out to find out. We measured the *Circle of Illumination* angles for over one hundred and fifty sites around the globe and plotted them. The results lead to the following discovery:

Among all the *Circles of Illumination* for these sites, we found two sets of sites whose ellipses are separated from each other at nearly equal intervals. In the first set the intervals divide the earth's circumference at the Equator in exactly twenty five segments. In turn each segment is divided in quarters whose circles are separated by a distance equal to 400.75 km. For their analysis we selected as starting point the menhir site which lies closest to the Equator near Aur Duri in Indonesia. We believe the designers would have selected this location as 'Prime Meridian'. The *Circles of Illumination's* intersection points with the Equator set the quarter hour markers of AN ANCIENT GLOBAL CLOCK.

The graphic below depicts the mechanics involved. Notice the symmetry of the *Circles of Illumination* and the reach of the sun at the Arctic and Antarctic Circles.

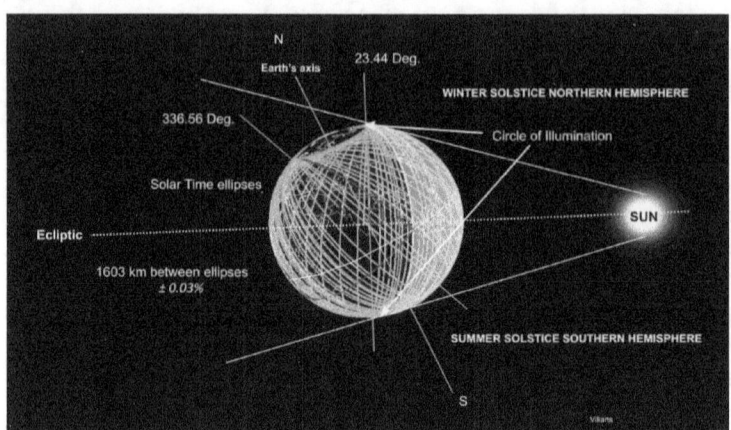

Graphic by the Author over a Google© earth -earth view

How these mechanics apply at a specific archaeological site can be best explained by looking at the Temple of Kalasasaya in Tiwanaku Bolivia. The Kalasasaya temple is part of another archaeological complex where the astronomical position of the sun at the summer solstice determined the design of its plazas and the positioning of its Menhirs, in particular the Ponce Monolith or Stela, and the Gates of the Sun and the Moon.

A common assumption regarding architectural positioning of monuments is that they were aligned North-South so that the structure would get the sun from the East at the Solstices. Although all those structures that are aligned this way do get the sun as prescribed, it is only in appearance; the actual alignment with the positioning of the sun at the culmination of the solstice is several degrees off 90°. How far off depends on the latitude the structure is located at and planetary fluctuations.

The Tiwanaku and Stonehenge monuments make this point quite aptly. At Tiwanaku the sun's position at the Summer solstice is at an angle of 114.53° and at Stonehenge it is at 50.61°; at neither site is the sun due 90° east, as it may appear to be to the naked eye. This fact is obvious at Tiwanaku. The Gates of the Sun and the Moon are aligned NS and EW respectively; the two structures are not on the same 90° line. However, both Gates are on the 114.53° line, the same as the Ponce Stela, such that at the culmination of the summer solstice these three structures are aligned with the sun. At Solar Time 0530 on December 21, 2014 the sun rising on the east aligned with the head of the Ponce Megalith, and both gates simultaneously at an angle of 114.54° [41]. At Stonehenge the geometry is similar; there is an alignment at sunrise at the culmination of the summer solstice between megalith 80 and the Heel Stone and megaliths 66 and 16, as shown in the graphic above. The graphics below, explain the earth's mechanics involved. The picture on the left shows the actual position of the sun early Dec. 2014. The cross-hatched lines show the *Circles of Illumination* at both Solstices and the systematic separation between them; each line corresponds to at least one archaeological site. The white circle line is the *Circle of Illumination* at the AurDuri Menhir; located at the Celestial Equator in Indonesia at the gray colored cross- hairs. Notice that an imaginary line connecting the centers of the earth and the sun is almost perpendicular to the white line as the summer solstice is approached in the southern hemisphere.

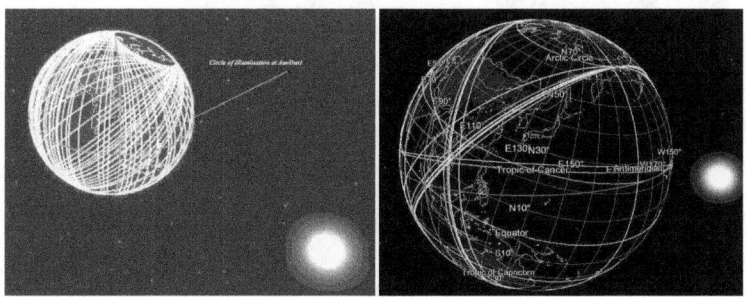

Graphics by the Author on Google© earth maps

The picture on the right shows the Sun's alignment at the summer solstice in the southern hemisphere and the *Circles of Illumination* (vertical lines) for Nasca, Machu Picchu, Tiwanaku at their antipode point. See a larger image on page 271

The horizontal lines show the direction of the sun perpendicular to the *Circles of illumination* of both Machu Picchu and Tiwanaku. In Tiwanaku, at the culmination of the summer solstice on December 21, 2014 at Solar Time 1200 the altitude angle was 83.12° and the Azimuth 180.00°, at sunrise 0530 hr the azimuth was 114.54 [41].

The close-up graphic below shows, the 90° east line the Kalasasaya Temple and the Sun and Moon Gates are aligned to. The horizontal line shows the actual path of the sun at the summer solstice.

THE ANCIENT CLOCK

The results for the first data set show that the Ancient Architects not only knew the astronomy of the earth axis' declination angle and the resulting equinoxes, but may have, also, known the earth's circumference in kilometers: 40075km. It appears they may have measured the days in terms of Sun Time.

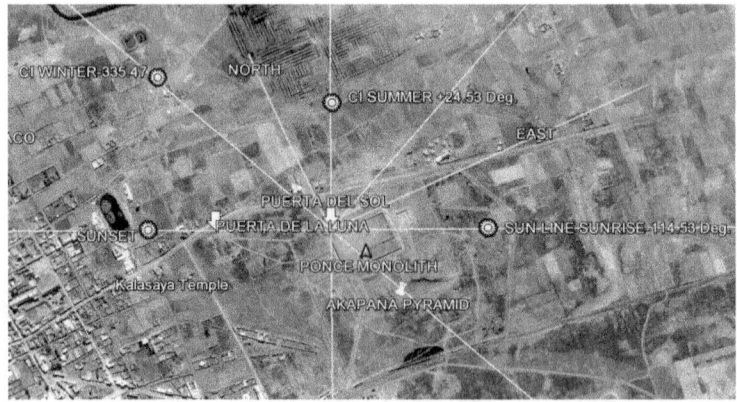

Graphics by the Author on Google© earth maps

They counted the daily cycles by dividing the earth's rotation time in 25 equal time spans (**Metric** hours, halves and quarters) - corresponding to 1603 km of earth's rotation for each hour, presumably assigning 13 hours of daylight and twelve hours at night for each day between the Vernal and Autumnal equinoxes - daylight savings time.

*This is a time scale not normally contemplated under any commonly known time measuring system. The 13 hour day may be related to the Aztec Trecena-thirteen days[42]. In a **mean solar** day the earth rotates slightly more than 360° due to the eccentricity of its orbit and precession.* This results in a rotational speed in Mean Solar hours of 1674.4km/h.

We used the mean solar hour to analyze the second data set. In both cases each quarter hour is dependent on the Circle of Illumination of at least one archaeological site which sets the position of the circle, i.e. the intersection with the Equator establishes a quarter hour tic mark.

For the 25 hour scheme, of the possible 100 quarter hours (400.75 km) we found markers for 66 (66%) of these quarters; with an average error of 0.48km. (0.29mi) or .12% error.

For the 24 mean solar hour, of the possible 96 quarter hours (418.6 km) we found markers for 78 (81%) of them; with an average error of ±0.50 km. (0.31mi) or .12% error.

The number of unique markers found, for each evaluation method are: 27 for the Mean Solar Hour and 13 for the Metric hour, the rest are shared by both methods.

Intuitively we had expected the time markers would be small constructions such as menhirs, cairns, circles, dolmens and glyphs. Of the total number of markers we found (91) made up the Ancient Clock, thirty eight are major structures or cities. The data is summarized starting on page 274.

SUMMARY

Early on in this study we discovered that the Nasca lines are directional paths to places where ancient civilizations lived. Some of the lines at these sites encode the earth's planetary mechanics and the encoding method is found in many structures around the world. In this study we found the location -latitude and longitude- of several structures represents an astronomical point on earth related to the mechanics of its orbit and rotation: the physics of time. Although most ancient societies developed time measuring methods with various degrees of accuracy, no one of the systems, thus far, reveal the knowledge of the planetary metrics that result in the mean solar day on earth (it may be encoded in the Aztec Calendar stone's or other glyphs). Also, none of them hint at the possibility of a global scheme, much less a coordinated design to form a Global Clock. This may substantiate the belief many researchers have that an ancient technologically advanced civilization existed on earth at a time earlier than our history.

CONCLUSION

We set out to find out why structures in archaeological sites were placed at their locations and what factors determined the selection of these locations. From the data analysis we conclude:

A prehistoric technologically advanced society left a legacy of a discreet group of archaeological sites which encode the earth's planetary mechanics. They encoded these mechanics by building into each structure's design the solar exposure measurements and cycles. The structure's geographical location and azimuth were selected to serve as geodetic markers for sun-earth positioning cycles phenomena. In most instances the structures, also, indicate the direction to other sites forming a coordinated network. The solar cycles recorded in the structures form a highly accurate Global Clock, with the precision of today's technology.

OUTLOOK

The results may open up a new line of inquiry if we begin to consider archaeological monuments as geodetic markers that record the historical astronomy of the earth. This instance is one example; we showed that the point selected for the ancient 'Prime Meridian' -the Aur Duri menhirs- is located 4.4km north of the Equator. This may indicate the menhirs may have been placed on the equator at an earlier time when the axial tilt was greater than the current 23.44°(accepted range 22.1° to 24.5) At the current axis drift of 10 cm/year southward along longitude 80°, antipode to Aur Duri (0.038°N 100.480°W); the menhirs could have been placed at the Equator 44,000 years ago, on its upward cycle at 180°! This number of years is immediately recognizable; it is the current earliest dated appearance of Homo sapiens in Africa at the Border cave in Swaziland[43] and in the UK at the Devon Cave[44].

The geographic position of archaeological structures may reveal the progression of the geophysical changes over time; taking into consideration Milankovitch cycles, in particular the oscillation of the earth's obliquity angle which, perhaps not coincidentally has a 41,000 year oscillation cycle[46]. At Nasca, the circle of illumination graphic shown earlier, illustrates a series of lines that could be a record of the shift in the earth axis' obliquity angle.

Another ther point of interest may be historical. Gunug Padang is a site containing numerous standing megaliths. It is exactly 15 minutes earlier than Aur Duri and its circle of illumination aligns in Colombia with El Infiernito, the Chicamocha Pictographs, the El Abra Caves and two lakes of legend: Iguaque and Guatavita. The first associated with the legend of Bachué 'The Creation of Mankind'-p.95- and the second with El Dorado. Is there a historical connection between Gunug Padang and El Infiernito as there may be between Aur Duri and the Hummingbird Pyramid?

Notes

CHAPTER 7

AREA 51 GLYPHS

The decision to include Area 51 glyphs in this study was difficult. We had to convince ourselves they were relevant. The great amount of lore and fanaticism that is associated with the region, was considered, could muddle any importance these glyphs could have. The area has been scarred with probably thousands of ballistic tests that can confuse the legitimacy of the so called 'Alien Glyphs'. Besides the Triangle, Fork, and Six point Star there are also, several lines, arcs and bull's eyes. Reluctantly we set out to evaluate each glyph using our established methodology: Follow the Lines!

The results of our evaluations can be seen in the graphic below. Our literature search, yielded the location of the main glyphs: The Triangle (A51 Triangle) with a bull's eye in it (37.63°N 116.85°W), a three pronged Fork (37.59° 116.91°W) and a Six point Star (37.40°N 116.87°W), not shown.

We identified the quarter circle, which has twelve arcs. The graphic below shows only two of them. The graphic is drawn to scale.

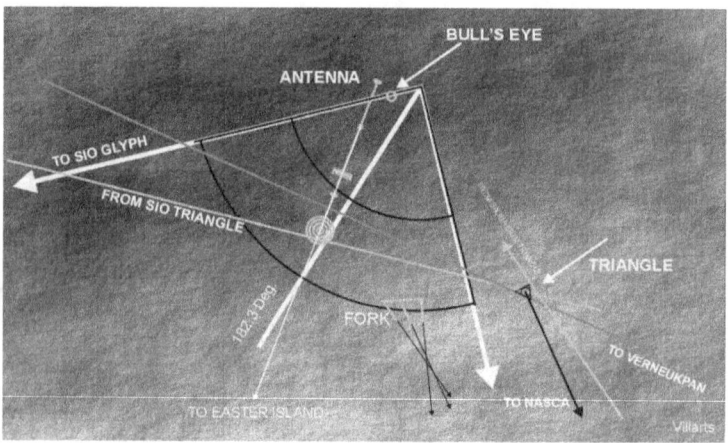

Northwest from the Fork we found what appears to be the representation of CB Antenna. No references were found that describe it, so we dubbed it "Antenna". It is shown above, next to line bisecting the angle. This glyph starts at five concentric circles; the largest has a diameter of 0.77 mile. The antenna's stem points north south at 172.58°and runs for 5.5 miles, ending at the apex of an inverted triangle, i.e. points southward.

On the way north at 1.14 miles there is a square bow-tie, as you'd find on a CB antenna (a lemniscates with the configuration similar to that found on the Sonora Stones). At 1.88 miles there is a cross member resembling a typical TV aerial antenna. At 3.69 mile there are two cross members. This strange glyph serves as the theme symbol for this chapter.

The importance this glyph brings to this study is the connection it makes with Easter Island. Following the Antenna's bearing, 4,490 miles south of it, on its path finds the Tautini Moai on the edge of the Rano Kau crater on the south end of Easter Island.

A line drawn nearly perpendicular to the antenna connects the base of the antenna - the five concentric circles center - with the center of the five inscribed circles in the Triangle, east of it, at an angle of 84.54°. The triangle's largest inscribed circle is exactly the size as the Antenna's third circle: 0.39 miles. The base of the A51 Triangle runs E at 90°, its apex is perpendicular to it at 180°.

Once again, we find the Celestial North is explicitly laid out next to another symbol or object having a slightly different bearing. In this case the antenna's 172.58°bearing which indicates purposeful design. This arrangement, together with the similarity in terrain characteristics of both the Area 51 and the Nasca regions, provided the encouragement we need to pursue and include the Area 51 glyphs in this study. We were glad we did, the results that were obtained added to the understanding of the global layout of ancient sites. At the site there are also, two Bull's Eyes which strengthen the argument for inclusion. The first one, less than a mile east of the top of the Antenna, consists of six concentric rings, the largest circle has a diameter of .19 miles; the same as the second ring at the Antenna's base. The second Bull's Eye is about 12 miles south. This Bull's eye has only two circles, the largest being the size of the third ring of the first Bull's Eye. These two Bull's Eyes are connected to a third Bull's Eye found at Nasca (14.64°S 75.17°W). The Nasca Bull's Eye has two circles, 0.01 and 0.02 miles in diameter. The circles for each figure can be nested in an almost perfect sequence from the smallest at Nasca to the largest at the base of the antenna. See graphic below. The arrows show the largest circle for each glyph. The three Bull's Eyes are connected by a line with a bearing of 324.53°. Recall, 324.86° is the bearing of one of Nasca's main lines.

The foregoing establishes the relationships between the Area 51 glyphs with Nasca and Easter Island. The table below shows the measurement's for easy review.

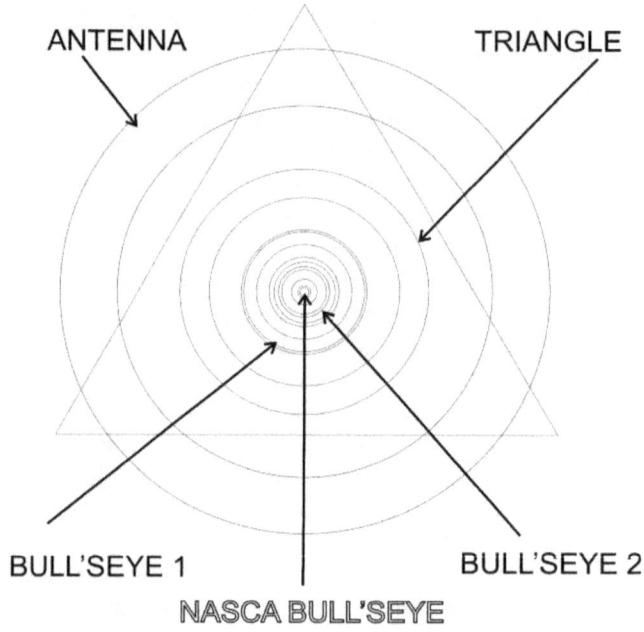

ANTENNA

TRIANGLE

BULL'SEYE 1

BULL'SEYE 2

NASCA BULL'SEYE

The Area 51 Triangle points north, it is an equilateral triangle of about .76 mile sides, which could almost totally, be inscribed within the Antenna's largest circle, which has a .77 mile diameter; as shown in the graphic above.

This triangle has three sister triangles; one at Salina Cruz, México, 1,979 miles away, which is smaller with half mile sides. However, this triangle is inside a larger triangle about the same size as A51: with .74 mile sides. The other two triangles are in the South Indian Ocean, 12,404 miles away. We'll discuss the SIO triangles in Chapter 8. The A51 triangle points north, while the Salina Cruz points south. The SIO triangles also point south.

CIRCLES DATA

SITE	COUNTRY LOCATION	Latitude	Longitude	Length miles	ANGLE°
ANDES CIRCLES		-12.42	-76.15		
Outer circle large diameter				560 ft.	
Inner circle large diameter				318 ft.	
Outer circle small diameter				515 ft.	
Inner circle small diameter				261 ft.	
ANTENNA	NEVADA	37.62	-117.00		
ANTENNA CIRCLES					
1st Circle Dia				0.02	
2nd Circle Dia				0.19	
3rd Circle Dia				0.39	
4th Circle Dia				0.59	
5th Circle Dia				0.77	
BULL'S EYE 1	NEVADA	37.56	-116.85		
NE line					16.39
1st Circle Dia				0.04	
2nd Circle Dia				0.08	
BULL'S EYE 2	NEVADA	37.69	-117.00		
NE line					16.46
1st Circle Dia				0.02	
2nd Circle Dia				0.04	
3rd Circle Dia				0.07	
4th Circle Dia				0.11	
5th Circle Dia				0.15	
6th Circle Dia				0.19	
TRIANGLE	NEVADA	37.62	-116.84		
TRIANGLE CIRCLES					
1st Circle Dia				0.07	
2nd Circle Dia				0.15	
3rd Circle Dia				0.23	
4th Circle Dia				0.31	
5th Circle Dia				0.39	
NASCA BULL'S EYE	PERU	-14.64	-75.17		
NE line					32.67
1st Circle Dia				0.01	
2nd Circle Dia				0.02	

The Salina Cruz triangle's western side has a heading of (324.45°), a familiar angle. A line run in that direction all the way to Area 51 ends at the vertex of the triangle found there. The Salina Cruz triangle also has a circle inscribed in it, unlike the A51 Triangle which has five. Both triangles are shown in the graphics below.

The connection with the South Indian Ocean Triangle-1 (T1-SIO) is similar. The East-West line of the T1-SIO triangle has a bearing of 88.90°. This line traces a great circle that passes through the center of the A51 Triangle and also, is nearly perpendicular to the Antenna (-6.32°). The T1-SIO triangle points southward at about 172°. The graphics below are drawn to scale.

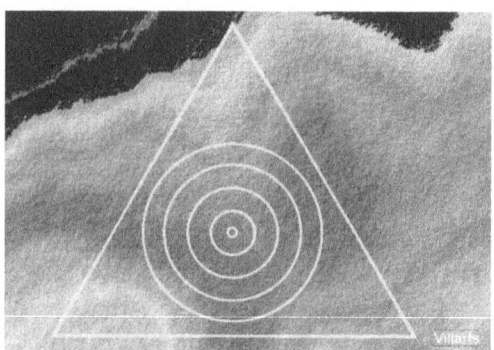

Area 51, Triangle & Salina cruz Triangle- Mexico

Following the Salina Cruz line on the opposite direction, down to South America, the great salt flats of Coipasa and Uyuni are found in Bolivia at the edge of the Sajama desert. On the same line, further south in Bolivia, the gold mine Soltera is encountered.

The A51 Triangle's eastern line has a bearing of 150.3°. Following this line, a series of gold mines are found in Nevada: Tiffin, Ingomar, Milford and Addison. In Arizona, the line finds the gold mines of Heart, Rattlesnake, Lord & Irish and Leiser Ray. In California it passes within sixteen miles from the Blythe Glyphs (33.79N 114.54W). The glyphs or intaglios are anthropomorphic figures inscribed on the ground, by apparently a technique similar to that used in the Nasca lines; the surface rocks are removed to expose the underlying dirt of a different color. These are discussed later in Chapter 9. Further south in Puerto Blanco, northern Sonora, México, more glyphs displaying anthropomorphic figures are found. Crossing the Pacific Ocean, the line passes about eighty miles east of Monte Verde in Chile where the oldest pre Clovis human remains in the Americas were discovered.

The A51 Triangle's east-west line, on its easterly direction finds Hovenweep. This Pueblos region is one of the archaeological sites whose location is triangulated: from Easter Island, Area 51, and Nasca. From easter Island by Ahu Vinapu Ceremonial Center 180°south of it, the A51 triangle 270° west of it and to the south, by Nasca with a line having a 141.87° bearing. Continuing east, the circle past Hovenweep reaches Africa. In South Africa the circle passes near Verneukpan, the home of hundreds of spirals, mentioned earlier. Most of the Verneukpan spirals are within lines arranged in an inverted 'V' shape and other smaller groups of various sizes outside them, the largest spiral being about .166 of a mile in diameter. Some call this location "the Nasca of Africa". Verneukpan will be discussed in detail in Chapter 8.

The A51 triangle's north-east line is, perhaps, of greater significance. On its northward direction, after crossing over the Arctic Circle, now southward bound it passes near the Ring of Brodgar and the mound of Maeshowe in Scotland. Cutting across Europe in Bosnia finds the Visoko Pyramid and the Athenian Acropolis. Over the Mediterranean Sea, it reaches northern Egypt near Abu Mena, Alexandria (30.84°N 29.67°W) and the Giza Pyramids complex.

Following the Nile River southward, further down it goes over Hermopolis (27.78°N 30.8°W), the Aysut Necropolis (27.16°N 31.17°W), then, it finds Luxor (25.72°N 32.66°W) and the Edfu-Horus Temple (24.8°N 32.87°W) on one side of the line, on the other the Hapschepsut Temple (24.8°N 32.03°W). Further south the Kom Omb Temple (24.45°N 32.93°W) and the Kalbasha Temple (23.96°N 32.87°W).

The A51 triangle is connected with Nasca with a line running parallel to the Nasca Great Arc, mentioned earlier, which on its way north passes fifteen miles east of the A51 Triangle. The line, starting at the center of the Triangle ends at its own line in Nasca, located at 14.68°S 75.1°W. This is the second connection between these two sites. Following the A-51 line north from Nasca it crosses over Tambo Colorado a mid-sized Inca city in Perú (13.70°S 75.82°W). This Inca city is more popularly known for the 'Band of Holes' a unique geo glyph discussed on page 14, located three miles west of it. In México the line crosses over Huamelulpan "The foundation of this ancient pre Hispanic city goes back to 400 BCE, it was an important urban center up to 800 CE; it is a good sample of the early Mixtec culture, called Ñuu Sa Na' or "Ancient People" (Ñuu Yata in the Mixteca Baja)" Wikipedia. Further north it finds Tenochtitlan (México City 19.44°N 99.13°W) the capital of the Aztec empire, and today one of the largest cities in the world. In the north side of the city, seven miles NW, it goes over the pyramid of Tenayuca. Further north it finds the Hidalgo Salt Lake (22.69°N 101.70°W) and the Bisbee Copper Mine in Arizona. The data for the Triangle is summarized below.

SITE	COUNTRY LOCATION		Latitude	Longitude	Length miles	ANGLE°
TRIANGLE SE LINE						150.30
Isla Serkik	USA	Island	37.63	-116.85		
Laiser Ray	Chile		-33.75	-80.43		
Lord & Irish	USA	Mine	35.00	-115.00		
Rattlesnake	USA	Mine	35.00	-115.00		
Heart	USA	Mine	35.23	-115.17		
Addison-Milford-Ingomar	USA	Mine	35.77	-115.48		
Tiffin	USA	Mine	35.77	-115.52		0.00
TRIANGLE E-se LINE						90.00
Hovenweep	USA	Pueblo	37.83	-109.07		
Vermeukpan -100+ Spirals	So. Africa	Glyphs	-90.10	21.10		0.00
TRIANGLE NE- LINE						210.14
				0.00		
Andranottsara-La saline	Madagascar	Salt Mining	-12.33	49.20		
Herto, Boun	Ethiopia	Homo Sapiens	10.29	40.53		
Kalabshah	Egypt	Temple	23.96	32.87		
Kom Ombo	Egypt	Temple	24.35	32.93		
Idfu	Egypt	Temple	24.98	32.87		
Luxor	Egypt	Temple	25.72	32.66		
Aysut Necropolis	Egypt	Burial	27.16	31.17		
Meryma-Amarna	Egypt	Burial	27.66	30.93		
Hermopolis	Egypt	Citadel	27.78	30.80		
Giza	Egypt	Pyramids	29.98	31.13		
Abu Mena	Egypt	Citadel	30.84	29.67		
Vienna	Austria	City	48.16	16.34		
Ring of Brodgard	Scotland	Megalith Circle	59.00	-3.23		
Bonnie Calire	Nevada,US	Gold mine	37.23	-117.13		
Salina Cruz City	Mexico	Triangle	16.24	-95.19	1979.00	142.00
Nasca	Peru	Line	-14.68	-75.10	0.11	324.45
SALINA CRUZ TRIANGLE	MEX		16.24	-95.19		
E-W line						89.83
S-E Line						142.25
N-E Line						25.63

SIX POINT STAR GLYPH

South from the A51 Triangle, at 183° is the six point Star. The 6P Star is made up of two opposing fairly symmetrical triangles, pointing north and south. This Star is connected with a 324.36° line from its center to a six point star at Nasca (14.7°S 75.1°W) 4,497 miles away.

The Nasca Star is about ten times smaller. The A51-Star also, has its own line on the Nasca plain, located at (14.78°S 75.24°W). Following this line south of Las Vegas, it goes over the El Dorado Canyon, Boriana, Total Wreck and El Tigre gold mines and the Cuarenta Casas cliff dwellings, shown below. Further south, in the State of Chihuahua it finds the petro glyphs of San Nicolás and the Cerro Blanco petro glyphs. In Zacatecas, it finds the Los Angeles silver mine.

Cuarenta Casas- Wikimedia Creative commons

The base of the triangle which points south, runs east west at 94.15°. A great circle drawn from it with this bearing, on its easterly direction, in northern New México passes thirty four miles north of Chaco Canyon and forty miles south of Hovenweep. The Pueblo settlements will be discussed in Chapter 11. In Louisiana the circle finds the mound at Poverty point thirty one miles south of it. This alignment connects three important indigenous cultures in North America. The east side of the same triangle runs north east at 36.25°. A line in this direction finds the Bonneville Salt Flats, in Utah and the Medicine Wheel in Wyoming.

Across the Atlantic, in Ireland it finds the Newgrange Mound.
In southern England it cuts across near the Avebury circle, the Bulford Stone and Stonehenge. In France, east of Paris the circle goes over the covered monoliths of Dampsmesnil. The western side of the triangle, in southern California, finds the Providence mine and the Old Woman Meteorite mine. The Six Point Star data is shown below.

SITE	COUNTRY LOCATION		Latitude	Longitude		Length miles	ANGLE°
SIX POINT STAR							
So. Triangle Top	USA		37.40	-116.87		0.13	94.15
W-line						0.12	154.42
E-Line						0.12	36.25
No. Triangle Top	USA					0.13	90.10
W-line						0.13	38.95
E-Line						0.11	153.97
SIX POINT STAR TO NAZCA							
Providence	PERU	LINE	-14.78	-75.23	4496.00	0.21	325.00
Old Woman	US	Mine	35.00	-115.50	183.63		154.42
Boneville Salt Flats	UT US	Salt	34.43	-115.20	225.25		154.42
Medicine Wheel	WY US		40.81	-113.73			36.25
Newgrange	Ireland	Dolmen					36.25
Avebury Ring	UK	Circle					36.25
Bulford Stones	UK						36.25
Stonehenge	UK	Circle					36.25
Poverty Point	LA US	Mound					94.15
Los angeles Mine	Mexico	Mine	22.48	-110.92			
Cerro Blanco	Mexico	Petroglyphs	24.95	-100.00			
San Nicolas de la Joya	Mexico	Petroglyphs	27.42	-106.22			
Cuarenta Casas	Mexico	Pueblo	29.55	-108.17			
El Tigre	Mexico	Mine	30.58	-109.22			
Total Wreck	USA	Mine	31.88	-110.58			
Boriana	USA	Mine	34.93	-113.92			
El Dorado Canyon	USA	Mine	34.70	-114.80			
Cuarenta Casas	Mexico	Pueblo	29.55	-107.58			
Star to Nasca Star	USA-PERU	Line		0.00		4497.00	324.36

THE FORK GLYPH

About five miles southwest of the A51 Triangle is the Fork glyph. This glyph consists of four lines. A top line with a bearing of about 54.38°, one and one half miles long. Starting from the western end a second line runs southward. This line is about a mile and a third long, and runs at an angle of 116°. A third line starts at the eastern end of the top line, it runs southeast at about 144°. A fourth line starting about the middle of the top line runs at an angle of about 127°. Continuing the top line northeast it passes at the cusp of the A51 triangle. In this direction the line crosses over the Bingham Canyon Copper mine, one of the largest in the world. Continuing the line as a great circle NE from the mine, it points to the uranium mines of: Sweetwater, Smith Ranch and Crow Butte, in Wyoming and Nebraska. Further up northeast it cuts through pre historic copper mines, carbon dated 5700 years, in Keweenaw Peninsula and Isle Royale MI, US. In Quebec Canada the line's path goes over the Selbaie Gold and copper mine. Over the Atlantic onto Africa, the line crosses 200 miles east of the Copper Belt.

Due to the differing angles the three prong lines crisscross on their way south. The eastern line points to Onavas and the Sonora Stones and the Alamo Dorado silver mine in México. The center line connects the now abandoned Hachita gold mine, a zinc mine in the Mexican state of Chihuahua, the petro glyphs at Puente Cuates, and the Maya City of Toniná in the state of Chiapas (6[th.] to 9[th.] Century); this city is at the end of the Great Arc. The southern line aligns with the Muyil pyramid, dated back to around 300BCE, in Quintana Roo state in eastern Yucatán, México. Crossing the Caribbean Sea into Colombia it finds the Chicamocha River canyon where petro glyphs are found at a location known as "la Mesa de los Santos", mentioned earlier. This region was the seat of the Guané Culture which had developed in pre-Columbian period and was present at the time of the Spaniards arrival in the XV century.

A notable point about this culture was their practice "to deform the cranium of babies, in both occipital and parietal directions"[45]

This practice was also found in the Paracas region of Perú (The location of the Candelabra), also, it was recently reported by Bryan Grenoble on the Huffington Post, similarly distorted skulls were discovered in the town of Onavas in the state of Sonora México, mentioned above. This site is located about forty miles north of the location of the 'Sonora Stones'. It may be recalled the Sonora Stones and Onavas are both mapped by the Ahu Hanga Kio'E line.

The fork's southern line, south of Onavas, connects the areas where deformed skulls had also been previously found of Sinaloa, Marismas Nacionales and Nayarit.

Archaeologist Cristina Garcia Moreno, project Director for the Onavas excavation, finds it surprising since these skull deformations had not been found this far north of Sinaloa, " The Onavas cemetery does not belong to Mesoamerican migrating groups, but, a sedentary type that developed locally...we are in the process of investigating whether at some time it contacted Mesoamerican groups and incorporated some of their cultural practices" http://terraeantiqvae.com

Many writers have claimed these elongated skulls are remains of extraterrestrial aliens, most notably by Author David Hatcher Childress. Skull deformation practices are widespread phenomena in many cultures; it is found not only in Mesoamerica and South America, but, in North West US, Europe, the Middle East, Turkey, Egypt and Bhutan.

> "An examination of the American monuments shows that the people represented were in the habit of flattening the skull by artificial means. The Greek and Roman writers had mentioned this practice, but it was long totally forgotten by the civilized world, until it was discovered, as an unheard-of wonder, to be the usage among the Carib Islanders, and several Indian tribes in North America. It was afterward found that the ancient Peruvians and Mexicans practiced this art: several flattened Peruvian skulls are depicted in Morton's "Crania Americana." It is still in use among the Flat-head Indians of the north-western part of the United States"[19].

AREA 51- FORK DATA

SITE	COUNTRY LOCATION	Latitude	Longitude	Length miles	ANGLE°
FORK					
North Line	Golden Eagle - Harquahala Gold mines-AlamoDorado Silver Mine	Arizona, US-Sonora, Mx		1.08	144.46
Center Line	Hachita Mining town- ZincMine, Chihuahua, Petroglyphs Coahuila, Tontina,Yuc.	New Mexico, US-Mex.		1.03	126.56
South Line	El Castillo, Yuc- Petroglyphs Chicamocha	Yucatan, Mex-Colombia		0.95	116.00
Top Line	Old Gold Mine, Uranium, Sweetwater, Smith Ranch, Crow Butte, Mine Selabie,Gold Copper,Zinc	NV, US-Que.Can		1.45	54.38

Elongated Skull's Connection

SITE				Length miles	ANGLE°
ONAVAS-CHICAMOCHA				2825.00	306.72
ONAVAS-PARACAS				3663.00	138.26
PARACAS-CHICAMOCHA				1429.00	9.45
EGYPTIAN AMARNA ART				7700.00	

The foregoing analysis of the area 51 glyphs' alignments reassured us these glyphs belong to a unique set of 'unexplained' archaeological sites: The Candelabra, Nasca, Easter Island and Verneukpan. Although the Area 51 glyphs studied make some connections to important archaeological sites, as Nasca and Easter Island do, it appears that the primary function of these glyphs was to map mineral resources. The glyphs, not only point to the usual resources we had already found through the other sites; Copper, Gold and Salt, but they added Uranium, which had been identified by an Ahu alignment. From the commonality of purpose and design perspective, the Area 51's glyphs are confirmed to have their place in this set.

The Candelabra lines pointed us in the direction of sources for copper, gold, salt and nitrate. It, also, pointed us in the direction of Nasca and leads us to the concept of associating the object's design lines with directions to other places. The Nasca's lines, sweep over the horizon 360° around, and point to the same mineral ores, but mostly, to notable archaeological historical sites. After we found mineral ores, guided by these two sites, we tentatively concluded this was a key to the logic of their existence. When the same ore sites were found, approached from different directions, following paths prescribed by different sites, it reinforced the conclusion we had reached earlier; the purpose of these sites and intent was to create a permanent record, a long lasting map of the locations they point to. Nasca pointed to Easter Island where we found Moai Ko Te Riku. Following this Moai's line of sight, added a third reference point for the location of the Congo-Zambia copper belt in Africa. Triangulation for a location adds indisputable evidence that ancient thinking beings, responsible for the design of these sites wanted to mark places like the Copper Belt for posterity. Before the A51 glyphs we had sufficient evidence for a conclusion; now Area 51 reaffirmed it.
The glyphs were found to poitnt to the Copper Belt, making it the

fourth site pointing to it. The Fork's top line on its easterly direction also passes about two hundred miles north east of it. That distance is outside our 69 mile radius constraint to be considered aligned, however, it is sufficiently close to be meaningful, taking into consideration that the same line connects three other important copper mining sites: Bingham, Isle Royale and Silbae. The distance between the Fork and the Copper Belt is nearly ten thousand miles. Later on we will discuss a fifth and sixth sites that make the same connection with the Copper Belt.

Notes

CHAPTER 8

Ocean Lines

Previously we had observed that a rectilinear
glyph or structure when it is not aligned with
the Cardinal points; its alignment in any other
direction, is probably aligned at that angle with
a purpose. That is, we have found that in those
cases where the alignment is at an angle
pointing in a direction other than the cardinal
points, a line drawn with that bearing, in most
instances it leads to important places, thus
confirming the alignment was set on purpose.
The Ahus of Easter Island are a good example,
as is the citadel of Teotihuacan in Mexico.

Neither the Teotihuacan Citadel nor its pyramids align
geographically with the cardinal points. The promenade "Avenida
de Los Muertos" The Avenue of the Dead, does not align with a
celestial body or a cardinal point; it has a bearing of about 15.15°.
Also, it has an unexpected layout; the larger pyramid of the Sun is
set to the side of the main promenade, while the smaller one, the

pyramid of the Moon heads up the arrangement. Some have ventured as far as comparing the layout with that of a computer circuit board[25]. For our study, the geographical alignment of the complex was the focus.

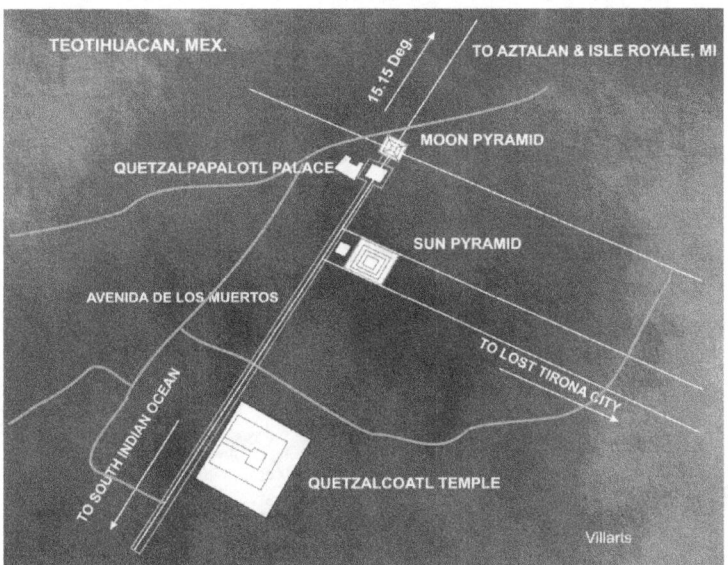

Archaeologist Marquina (1951)[3] believed the pyramid of the Sun, whose axis is perpendicular to the promenade was arranged so to face the sunset. During the summer solstice on June 22, 2014 the sun's azimuth at sunrise was 65.07°, thus the sun illuminated the eastern diagonal side of the pyramid. In winter the sun sets at about 80° to the western side of the pyramid.

Earlier, on page 108, we had explained that lines drawn on either side of the Pyramid of the Sun, with a bearing of 106.14°, nearly perpendicular to the promenade, lead to northern Colombia where the two lines encompass the length of the steps at the Lost City of Tyrona. The alignment of the pyramid with Tyrona, makes Tyrona's location geometrically related, or dependent on the promenade's and the pyramid's bearing.

Following the promenade's line southward around the globe to the south Indian Ocean, passes over northern Antarctica. This line, together with a line with a bearing of 224.71° from a megalith arrangement in the El Infiernito, which goes southwest through Easter Island and also follows the Ahu Tu'U Tahai's direction, reaches the south Indian Ocean south of Australia. Both these lines between southern Australia and Antarctica encompass -like a parenthesis- the Indo-Australian Plate southern ridge line.

THE SOUTH INDIAN OCEAN RIDGE LINE 'GLYPH'
This ridge line runs east-west for about three thousand six hundred miles between Australia and Antarctica. Many of these deep ocean lines have been reported by oceanographers and mapped with Google[30] Carrying out a close examination of the ocean, at an eye altitude of about 250 miles, we were able to see the ridge line. Intrigued by the South Indian Ocean (SIO) ridge line's very peculiar

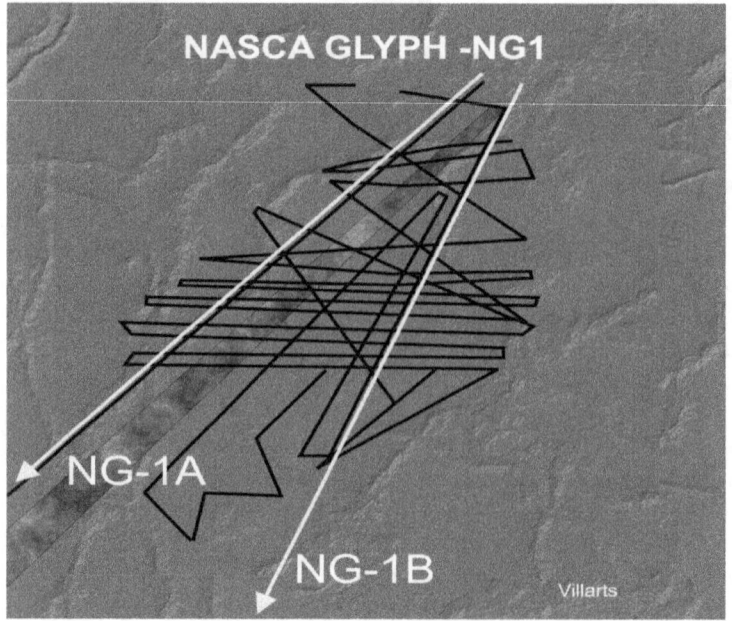

geometric pattern, we proceeded to trace it. A graphic of a portion of the glyph appears on the glyph page at the beginning of the book and at the beginning of this chapter, p 175. Our tracings are similar to those reported but clearer than the computer generated ones. We call this ridge line The Great South Indian Ocean Glyph (SIO Glyph). In the study, this was the first close encounter with geology and ridge lines. Some sections of the glyph reminded we had seen similar patterns before, somewhere on earth at another place: at Nasca.

At Nasca we had found line glyphs depicting the same graphic technique; a single line tracing, zigzagging, turning squarely side to side and eventually doubling back onto it returning to the beginning of the drawing. The graphic on the previous page, is one of several similar glyphs found at Nasca (14.64°S. 75.06°W), this one resembles a segment of the SIO Glyph. The graphic on the following page is the fragment of the SIO glyph that compares with the Nasca graphic; it is found at 40.96°S 78.67°E.

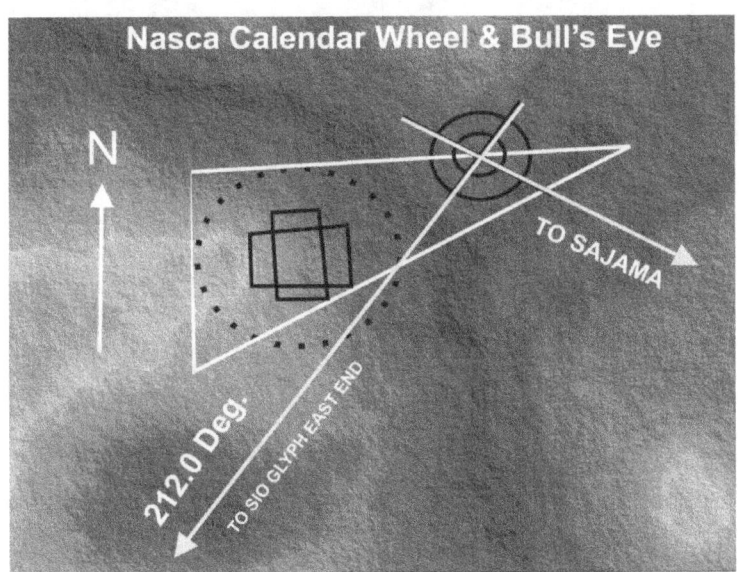

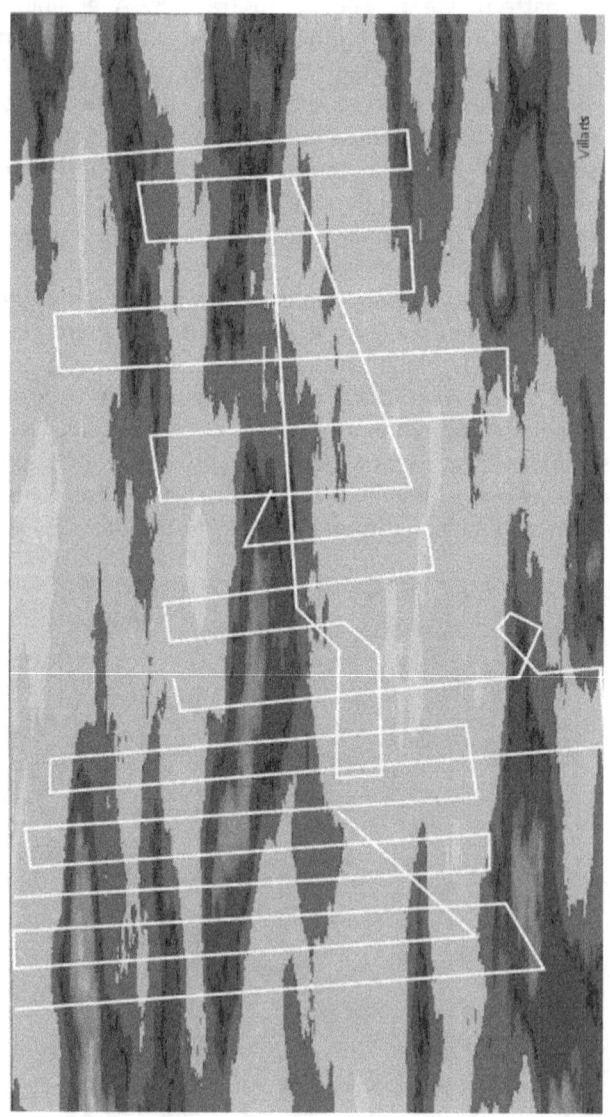

South Indian Ocean Ridge (Glyph), Detail

What causes ridge lines to develop in such patterns and of this magnitude (the segment shown is about 170x140 miles in size), we don't know. The only thing similar to it, in our experience would be dry clay; it cracks forming slabs that approximate this arrangement. The earth during its solidification process would have had similar physical characteristics? Why the folding serpentine arrangement?

The answer will remain a mystery. We had hoped these patterns were a figment of the technology, but the oceanic research[30] evidence shows otherwise; it is compelling. Google©earth plate boundary models are available as .kmz files.

What we discovered is that the authors of the Nasca lines knew about these patterns and enshrined them in their glyphs. Support for this claim was found following the lines. The eastern and western lines of glyph NG 1; NG1-A and NG1-B shown on p. 176, encompass 1100 miles of the Great SIO Glyph.

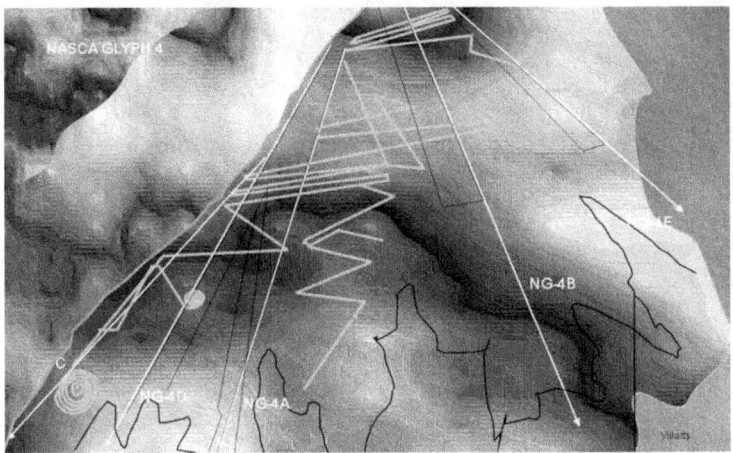

The second glyph NG4, above, has five lines A through E which, all point to the SIO Glyph; lines NG4C thru NG4E encompass 3,295 miles of its length. Line NG-4E points to the western most point of the SIO Glyph. The SIO Glyph's easternmost point, in Tasmania, is pointed to by the Nasca Bull's Eye (NBEG) glyph, P.177.

The total length of the SIO Glyph is 3,675 miles encompassed in its totality by the NBEG1 and NG4E lines.

North of the Area 51 Triangle, there is a Nasca like Polygon that runs north-south. About 3/4 of the way up is the center of a circle 9 miles in diameter, which contains another 11 concentric circles for a total of 12 circles. The actual glyph is only -exactly- a quarter of a circle. The quarter circle has a radius marked in the middle at 45°. That line is 2.3° off the 180° N-S line. Its western line, therefore, points 227.3° SW. Again, this 2.3° deviation from 180° south displays intentional design; set to follow the Indo-Australia ridge line -the SIO Glyph. This line runs parallel to the SIO Glyph, over 12,000 miles away! The line runs from top to bottom of the SIO glyph graphic on page 184. The picture below shows the quarter circle and the start of the line pointing westward to the SIO glyph.

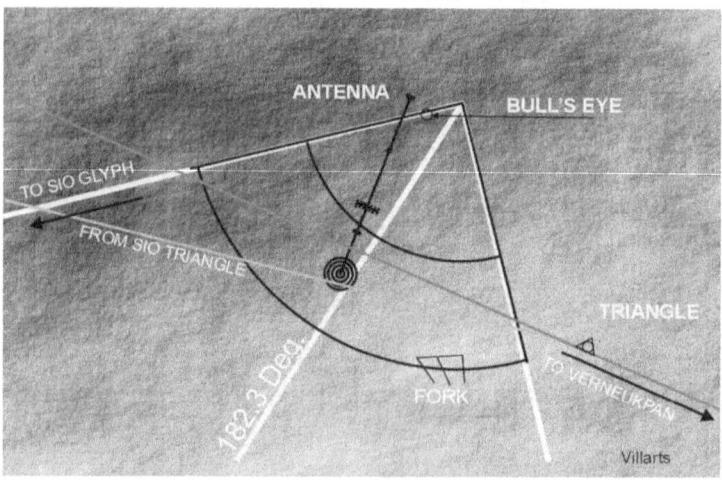

THE SIO GLYPH

In the picture on page 184, the short gray and white arc lines are the 'parenthesis' formed by the Teotihuacan and El Infiernito lines.

The 'Great Arc' traced which connects the Caracol with Göbekli runs parallel to a 1,500 mile line of the SIO Glyph- top left. It is shown as a dotted line on the graphic. The upper horizontal line on the eastern side of Tasmania is the Nasca Bull's Eye line. The lower horizontal line at the Glyph's end is the NG4E line. Following that line south of the SIO Glyph, 430 miles away, it passes over the western vertex (SIO-V3) of the SIO Triangle 1. This is the third triangle we had mentioned earlier as the sister of the A51 and Salina Cruz triangles. The line connecting the vertices SIO-V3 and SIO-V1 runs north east at 89.65°, a line traced following this direction connects the T1-SIO triangle with the A51 triangle at its center. It is shown as a light gray line at the bottom of the graphic on page 185. Continuing the line westward onto southern Africa, it passes seventy nine miles NE of Taung where Australopithecus Africanus -Tung Child- was found in 1924. We find that it is common place for the ocean glyphs geometries to point in the direction of places and sites on the continents. An important example of this is the third horizontal line up from the glyph's western end (bottom of the picture on the right); it is an extension of the main ridge line; it is labeled: 'To Sajama Glyphs'.

This line is significant in that it points to Lake Titicaca and the Chincana ruins in the Isla del Sol in the lake and passes seventeen miles south of the mysterious door of Amaru-Meru Hayu Marka; locally believed to be a portal to another dimension. South from the ruins, the line passes twenty six miles east of the pre-Inca City of Tihuanaco, the Temple of Kalasasaya and the Puma Punku Citadel. Further south it finds the ruins of La Fortaleza in Rosario, Bolivia and the Aymara burial Chullpas and the Lines of Sajama.

The Sajama lines which appear to have been created using the same ground scrapping technique used in Nasca, cover an area immensely larger than the Nasca lines. Project Sajama, of the Landmarks Foundation reported a total area covered of 22,225 square kilometers, approximately fifteen times the size of Nasca.

The line from the glyph cuts the Sajama Altiplano east of the Sajama peak (elevation 20,423 feet) and through what appears to be a core of the Sajama Lines. In the area along the line, within the region north of the Salar de Coipasa we cataloged 107 (location is given in the appendix) Line Nodes and three circular mounds cut up in concentric segments, ranging from 100-300ft in height. The mounds are arranged in an equilateral triangle of 1.2 kilometers from their center points. The node points are the starting point of lines that spread out in all directions, similar to the ones found at Nasca. Some investigators believe these lines precede the Nasca lines. Investigators at the University of Pennsylvania have mapped some of the lines, west of the Sajama snow peak, under a former project named Tierra Sajama http://cml.upenn.edu/tierrasajama/Sajama(now removed).

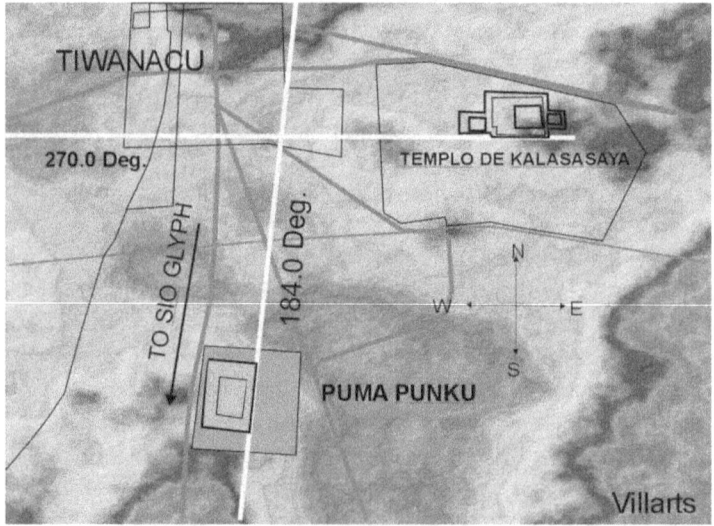

The city of Tihuanaco had been estimated to date back to 15,000 BC by Arthur Posnasky, whose calculations were dismissed as invalid by most scholars, who believe it dates back to between 300 BC and 1000 AD. The Temple of Kalasasaya, in Tihuanaco, is aligned with the cardinal points: 0°N-180°S. A short distance (.67mi) southwest of it is the Puma Punku Citadel. This citadel is aligned nearly 4.0°N-184.0°S. A line drawn in this direction from the Citadel, reaches the SIO glyph at the east end of the fracture zone. Apparently, the citadels' builders, also, knew of the ridge's existence, but, at a time earlier than the Nasca line authors did. As mentioned, the Tihuanaco Empire is commonly believed to pre date the Inca Empire. Was Mr. Posnasky correct in his dating calculations? The alignment of the Temple and Citadel is shown in the picture above. Again, Puma Punku's misalignment off 180° is shown to be intentional, the difference is pointed out by the Kalasasaya temple.

MORE OCEAN LINES

There are other oceans glyphs which have connections with some of the landmarks we have discussed, or connect to other ocean glyphs. We start with the latter type since they are related to the SIO glyph. Staring atop left on the picture on the next page, there is a line that passes through Tasmania in a west to south east (diagonal in the graphic) direction across the graphic. This line originates at an ocean glyph in the Pacific Ocean just off the coast of Oregon, P.186. This glyph PNA-1 has a right triangle paper clip shape. The line in question, corresponds to the triangle's hypotenuse, it starts at (39.8°N 134.01°W) and connects the two glyphs. The triangle's NW-SE line, the gray line starting at PNA-1 in the world map on page 186, cuts through the Baja Peninsula goes through México connecting the Tula pyramid with Teotihuacan, the Zapatera petro glyphs in Nicaragua and El Abra Tomb, just south of El Infiernito in Colombia.

In the Atlantic Ocean east of the Bahamas there is another glyph, A-G-1, consisting of an east-west line 86.59° over 1,800 miles long, starts off the coast of South Carolina and reaches the Atlantis Fracture zone. This line is intersected by a pair of lines southward bound with a 186.3° bearing, two hundred and thirty eight miles from its starting point. These lines are eight miles apart; both lines run southward for about 530 miles. Following one of the lines over Antarctica, it reaches the SIO Glyph 1,700 miles from its western end. This line - not shown- crosses towards the middle on the SIO glyph, immediately next to the El Infirnito line, which

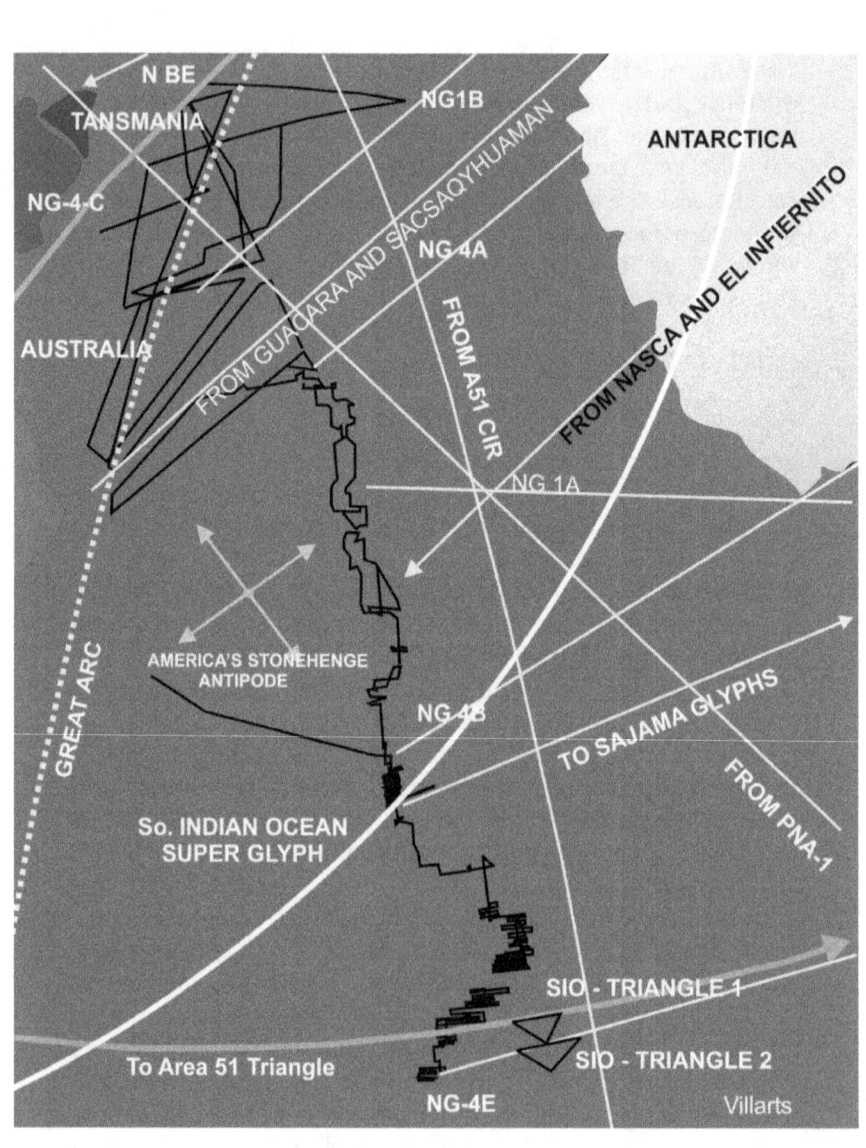

N BE
TANSMANIA
NG-4-C
AUSTRALIA
NG1B
ANTARCTICA
NG 4A
FROM GUACARA AND SACSAQYHUAMAN
FROM A51 CIR
FROM NASCA AND EL INFIERNITO
NG 1A
AMERICA'S STONEHENGE
ANTIPODE
NG 4B
GREAT ARC
TO SAJAMA GLYPHS
FROM PNA-1
So. INDIAN OCEAN
SUPER GLYPH
SIO - TRIANGLE 1
SIO - TRIANGLE 2
To Area 51 Triangle
NG-4E
Villarts

connects through its own line at Nasca -black lettering.

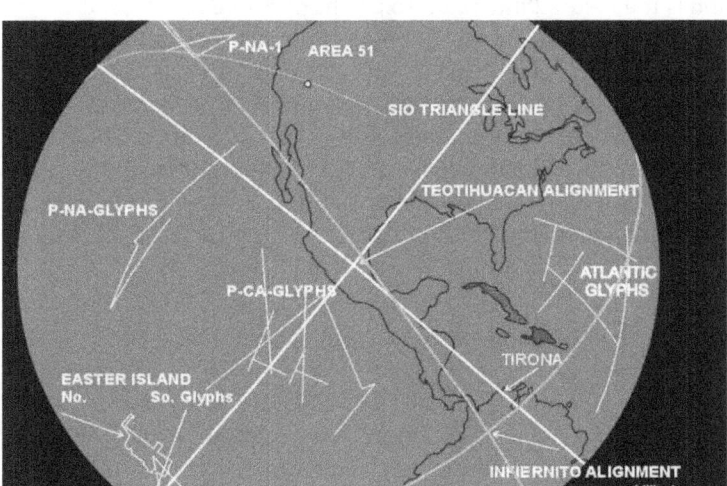

Following the Atlantic glyph A-G-1 parallel lines south, at about 372 miles from their starting point they cross another ridge line A-G-5 that runs NW-SE. A line drawn over it, on its SE direction of 111.38° reaches the SIO glyph at its eastern end.

On its NW direction it passes near the King's crossing, Raffman and Poverty Point mounds. In Australia the line goes over a region where nodes are found, similar to those found at Sajama. No reference to them was found. The Atlantic & Pacific glyphs are shown above.   Another ocean line starting at 30.78°N 65.76°W, almost perpendicular to the Miami-San Juan line of the Bermuda Triangle, also reaches the SIO Glyph following a bearing of 212.7°. Before reaching the SIO glyph it passes over the southern end of the Easter Island Ocean Glyph.

GUACARA

North of Guacara, in Venezuela (10.3°N 67.89°W), there is another less known menhir alignment in South America. This alignment has a bearing of 190°. Immediately south, in this direction is the Piedra Pintada petro glyphs park in Guacara. These are considered among the most important archaeological finds in Venezuela. A line drawn at the menhir alignment's angle reaches the SIO glyph between lines NG-B1 and NG-4A. The purposeful alignment of this menhir is underscored by the fact that it points at the SIO glyph connecting through Sacsayhuaman Cusco, the capital of the Inca Empire. Following the line down into Perú it finds Sacsayhuaman, the Inca Jail and the Toro Muerto petro glyphs.

Across from the Panamanian isthmus, in the Pacific Ocean there is another set of line glyphs, labeled in the picture on page 185 as P-CA-Glyphs. There are three of these glyphs. The furthest out to sea are P-CA-G-1 and -2 both resemble the letter A and their designs are almost duplicates of each other. Connecting the two glyphs is glyph P-CA-3. This line has a zigzag shape. It is composed of one NE line that turns SE before reaches the coast of México, travels over eight hundred miles until it reaches Guatemala, where it turns towards its coast. It runs for over one hundred miles and turns sharply south for another five hundred miles. Starting with the P-CA-1 glyph we traced its north and east lines. The north bound line cuts through Nayarit and Marismas, the two locations where deformed elongated skulls were found. In northern US it passes about one hundred miles east of the Medicine Wheel. The eastbound line reaches the El Abra Tomb in Colombia. Graph on p. 188.

Glyph P-CA-2 is east of P-CA-1, its NW, and east lines were also traced. The NW line, in México, connects: Nayarit, Marismas, Sinaloa, the Sonora stones and Onavas. Elongated skulls have been found in all these areas. Before reaching the US it goes over the Cananea Copper Mine. The northbound line connects the Palma Sola archaeological site and the Huamango Pyramids. The eastbound line runs parallel to P-CA-1's east line, in Colombia, thirty miles north of El Abra, finds the Salt Mines at Zipaquirá. We traced all the lines of glyph P-CA-3. The NE line crosses the Mexican coast line at Palma Sola archaeological site in Acapulco. Across México, before reaching the gulf, it finds the Defensa and El Tjin archaeological sites.   Across the Gulf of México into Louisiana the line passes, less than forty miles east of Poverty Point. The SE line tracks into Perú connecting the Inca cities of Cusco and Sacsayhuaman and the surrounding archaeological sites of Harucondo, Sondor, and Apusoconta. Further south in this direction it tracks the length of Lake Titicaca, passing by the Chincana Ruins on Isla Del Sol, then, it passes twenty five miles south of Tihuanaco and Puma Punku, further south it finds Fortaleza Rosario and the glyphs of the Sajama region, which are, in turn, connected to the SIO Glyph, as was discussed earlier.

PACIFIC OCEAN GLYPHS EXTENDED OVER LAND

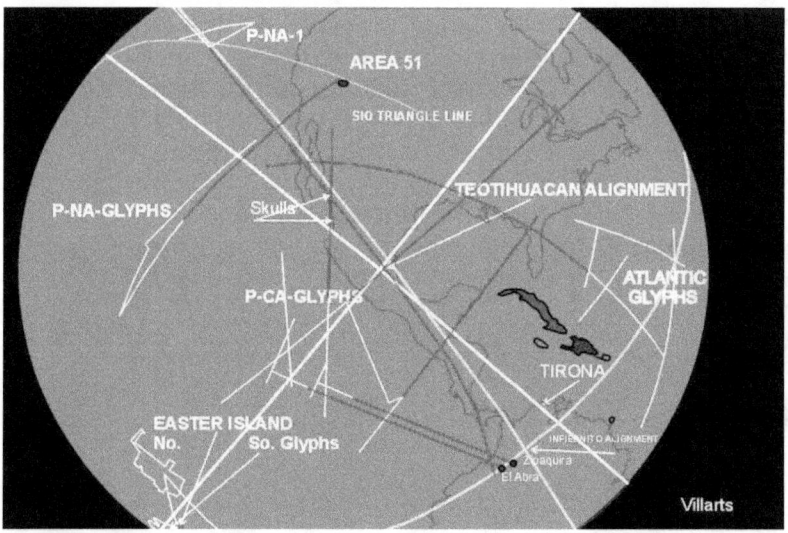

The third line of P-CA-3 glyph in its NE direction cuts through Nicaragua, connecting the archaeological sites of Zapatera and Sulutara. Across onto the Atlantic Ocean, it finds the glyphs off the South Carolina and Florida Coasts, discussed earlier. The Pacific

South American Glyph- P-SA-1 is a two line glyph that starts in the Pacific Ocean east of the Toroko Seamount Chain, forming a 'V' shape with a line from the NW, 5,300 miles long, having a bearing of 113.61°. A line extension from the first line, points to the Arequipa region of Perú at an angle 91.11° where it finds the Cerro Verde Copper mine. West from there, it reaches the southern side of Lake Titicaca, fifteen miles from Tiwanaco and Puma Punku. The second line's extension, points to the Sajama region, and becomes the south west boundary of the region, north of the Salares of Coipasa and Uyuni. The 'v' shaped glyph is shown below pointing to the South American coast line.

OTHER GLYPHS:

The graphic below shows several ridge lines. the one in the west; we have named the South Pacific Ocean Glyph. In the east are the Easter Island glyphs, which are located north and south of it. The South Pacific Ocean (SPO) glyph and the Southern Easter Island glyph are discussed staring on page 194. The SPO glyph's relationship to the pyramids of Giza is discussed in chapter 10.

Easter Island and South Pacific Ocean Glyphs

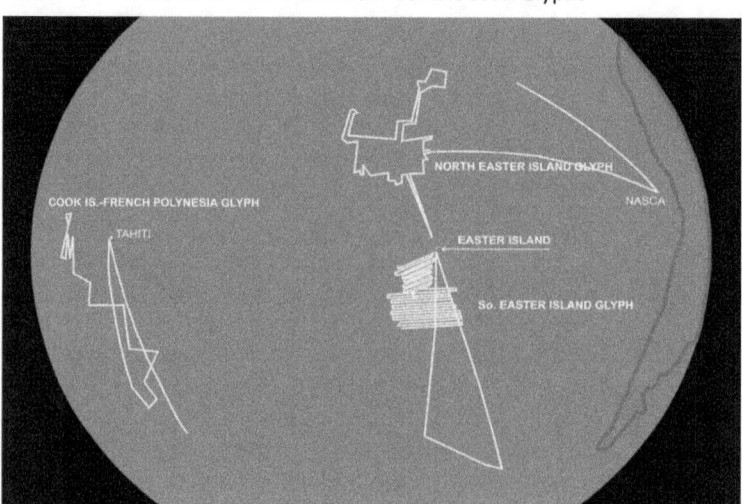

THE VERNEUKPAN SPIRALS

Verneukpan, South Africa (30.01°S 21.1°E), shown below, was mentioned earlier in the Symbols section, is the location where a vast number of spirals are found. One of the largest is about 571 feet in diameter; most others are between 150-200 feet in diameter. The graphic shows the inverted "V" bisected through its vertice, in which most spirals are enclosed.

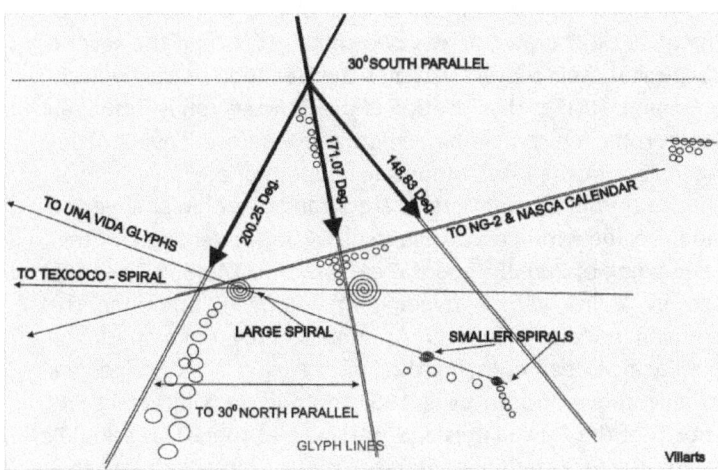

The three lines which form it are part of the glyph as is the line that intersects these in a southeasterly direction. The rows of spirals and boundry lines marked on the ground point in directions to archaeological sites in the manner we found at Nasca. The largest spiral in the world is the Texcoco Spiral found in in México City. p.21. A great circle starting at the center of the spiral in a southeast direction across the Atlantic, in South Africa finds the Verneukpan spirals, where it crosses by the two largest spirals, of the hundreds found there. Before leving Mexico the circle goes over San Lorenzo Tenochtitlan the site where the great Olmec stone heads are found. The connection between spirals and anthropomorphic figures is widespread particularly in petro glyph art. Notable in stelae are the spiral glyphs on the back of

Moais mentioned earlier. The same great circle on its northwest direction of 293.51° finds the circular pyramid of Peralta and near the Pacific Ocean coast goes through the archaeological region of Nayarit where elongated skuls were found. In Nayarit near Atacán it finds the Galana Stone with an engraved spiral on it. The spiral is notable due to the radii engaved on it; the only other spiral with radii we have found is the Texcoco spiral.

At Verneukpan some of the spiral alignments lead in the direction of other well known archaeological sites. A line with a northwest hedding of 291.62° drawn across two small spirals and the second largest spiral at their centers, reaches the Una Vida petro glyphs in Chaco Canyon, NM; at this location is the Chaco glyph is found, an atrophomorphic figure holding a spiral p.25. Before leaving Africa the line goes over the Black Mountain copper mine.

Most of the spirals, as mentioned, are arranged within an inverted 'V' shaped region within a 1.5 mile radius. The western side of the V has a heading of 200.25° and the eastern side 148.83°. The 'V' is bisected by a line with a heading of 171.07°. Continuing the western and middle lines over Antarctica, they encompass the western Pacific Ocean encasing the South Pacific Ocean ridge line. See graphic above and on page 198. The two lines intersect 540 miles north of the Hawaiian Island of Kaua'i, at the 30° N parallel. At this point where the lines intersect there is an ocean line that marks this parallel; it runs west until it reaches the Izu-Ogasawara Trench east of Japan.   Verneukpan's inverted V's cusp is on the 30th South Parallel in Africa. The Verneukpan 'V' arrangement is a segment of two intersecting circles whose intersection points connect the south and the north, 30°Parallels. Connecting the circles' intersection points describes another great circle, tangential to these prallels.

A circle following the direction of Ahu Akahanga, it may be recalled connects the same two parallels. These two circles' positions at their corresponding 30°N 30°S tangential points are skewed by 20°in longitude. Verneukpan is located ten degrees west of Giza, the Verneukpan NE line at 20.25° bisects the Ahu Akahanga great circle ten degrees east of Giza at the point where

the circle reaches its maximum latitude of 30.398°N 41.898°E.
A line running directly north of Verneukpan crosses the north 30th parallel at 21°E; the Pyramid of Keops is at 31°E and the intersection of the great circle is at 41°E. These appear to be purposefully designed alignments that set a geometric relationship between the pyramid of Cheops and the Verneukpan spiral arrangement.

Another line which cuts across the top of the inverted 'V' with a 254.2° heading, part of the design, which aligns with a row of spirals and the second largest spiral, when continued on its westward direction reaches Nasca. See graphic on page 189.

Before reaching Nasca it cuts through the northern Sajama desert and over the volcano. In Nasca it passes near the Nasca Glyph-2 (NG-2) and the Calendar Wheel. At this site there is also the Nasca Bull's Eye.The 254.2° line coincides with one of the Bull's Eye lines which cut across its center; from there it continues on to the north Pacific Ocean, where it reaches the vertex of the inverted 'V' lines on the 30°N parallel. This point is the antipode point to Verneukpan. Three miles west of this point a ridge line begins which is made up of two lines that converge at a point one degree north of the parallel, about 69 miles. At the point where they converge we find an ocean glyph we have dubbed; 'The Hawaiian StarAngel', with its center at (31°N 159°W).

A graphic of the glyph's main lines is shown on page 192. The actual 'glyph' is quite detailed. The glyph has a line at the center of the star with a bearing of 52.21°, it runs across the US and through America's Stonehenge in New Hampshire. Before getting there, in the State of Washington the line crosses a site on the upper Columbia River where petro glyphs are found, mentioned earlier. These petro glyphs are of unique significance because their design are claimed to match that of plasma bands artificially created to mimic high energy bands hitting the earth; which are also claimed to correspond with the 56 post points at Stonehenge and to the Calendar Wheel stones in Nasca[27]. Further east the circle following the line's heading crosses the city of Kennewick where a 9,000 year old man of Polynesian ancestry was found on the shores of the Columbia River (46.22°N, 119.14°W)[21].

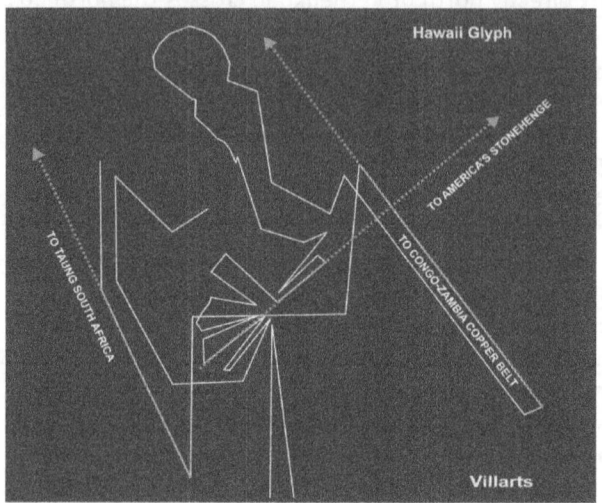

"The massive stone heads on Easter Island don't stare out to sea, but perhaps they should. Residents of what's also known as Rapa Nui sailed back and forth to the Americas hundreds of years before European explorers first reached the isolated Polynesian island in 1722, a DNA study suggests". Bruce bower -12:22pm, october 24, 2014, Sciencenews.org

In Michigan the line traces the copper mining sites of Isle Royale on Lake Superior.   Another one of the glyph's lines runs at 328.68°, cuts through Africa passing about one hundred miles from the Herto Buri peninsula, Olduvai Gorge, and Taung in South Africa. A third line with a bearing of 340.36° cuts through the Congo -Zambia Copper Belt. This is the seventh connection to this site and the second copper connection for this glyph and the fifth for Isle Royale copper. The geometry and locatin of these ridge lines is hard to explain what their origin might be.

Off the coast of Japan near the island of Yonaguni is the underwater pyramid (24.4°N 123.01°W) found in 1989, mentioned before, which aligns with Stonehenge and the Abbas Giant. This pyramid has a prominent Nasca line at (14.78°S 75.18°W) which is part of a trapezoidal figure. The line on its path to Nasca passes thirty three miles north of the Hawaii StarAngel. The StarAngel glyph is enclosed by Verneukpan's inverted 'V' lines on their northward direction: 20.25° and 351.07°.

THE SOUTH PACIFIC OCEAN GLYPH

The wedge-shaped region of the globe with vertices at Verneukpan and on the 30th Parallel north of Kaua'i encloses the South Pacific Ocean 'ridge line or ridge' (Glyph). The SPO glyph starts at a formation of lines arranged in oblong shapes (70.49°S 152.91°E) north of Antarctica.

The line runs north 0°, for about 3,610 miles (~6000km) and coincides with the150° W meridian. It runs parallel at about 1° west of the great circle generated by the Pyramids at Giza north alignment of 0° (This relationship is discussed at length in Chapter 10), until it reaches within eighty miles south of the island of Tahiti; from there it returns southward for about 1250 miles, then it returns back north in a zigzagging manner similar to that of the SIO glyph.

At the end of the glyph (10.92°S 158.48°W), it connects with an ocean line running NW at 42.8°, which if continued overland, at Baja California and onto New México, it crosses over the Fajada Butte in Chaco Canyon. The graphic on page 199 shows the South Pacific Ocean glyph detail. The SPO glyph, like the SIO Glyph, was also known to the Authors of both; the Nasca lines and the Verneukpan's spirals. In Nasca they engraved the ocean glyph's image in, both, the Geo Glyph 2 (NG-2) shown below and as part of the Monkey glyph. Both glyphs designs contain design elements which closely simulate the South Pacific Ocean Glyph. The Monkey glyph, is shown on page 195. The monkey graphic depicts the Pacific glyph more accurately than the NG-2 graphic below.

Both designs point directly to the Pacific Ocean Glyph. Starting with the NG-2 glyph; its design has a spiral within its zigzagging lines, the spiral's end line cuts across the glyph at 275°. This line reaches the top end of the SPO glyph. This is where the 42.8° line starts which goes to the Fajada Butte. The middle line of the NG-2 zigzag points at 240.39°, in the direction of the Pacific glyph. The line reaches it in the middle of its own zigzagging line.

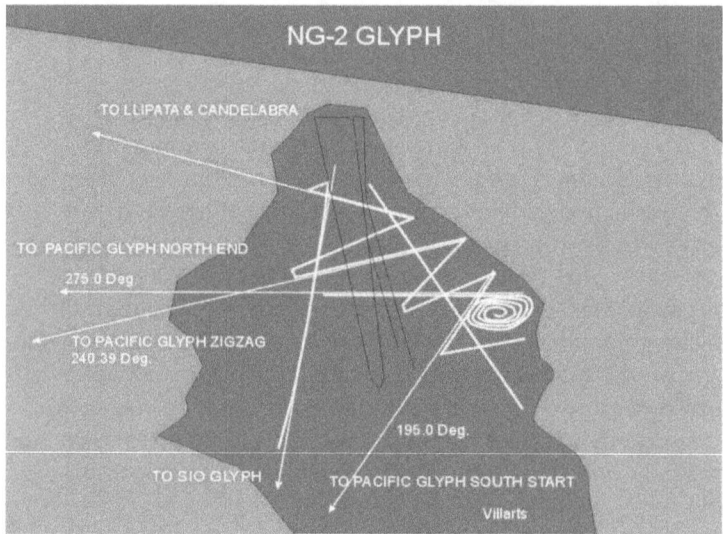

A second line of NG-2 which encloses the spiral, points SW at 195.0°. This line leads to the beginning of the SPO at its southernmost point north of Antarctica. The upper line of the NG-2 glyph points NW; in that direction north of the Nasca plain, it crosses the site where the famous 'family' glyphs are carved on the side of a mountain at Llipata Perú, shown below.

The headdress on the glyphs on the right and left of the picture resemble the Una Vida, the Chelly Canyon glyphs, the Newspaper glyphs and the Tirona golden warrior's helmet. p.13.   The Pacific Ocean glyph is also enclosed by the Track at the Acropolis in the Greek Island of Rhodes. The track runs NE at 4.56°. A line continued north at this angle, cuts through the SPO Glyph lengthwise from north to south, the gray line p. 198. A line drawn perpendicular to the track 274.56° going west reaches the glyph at its southern end; these two lines encase the glyph in its totality. The Track's east-west line cuts through Perú at the site of the archaeological ruins of Cañoncillo and the Ventanillas burial sites of Arascorge and Atoshaico. These burial sites are strikingly similar to the Necropolis at Aysut in Egypt (27.16°N 31.17°E). The Cañoncillo site is estimated was established 2,440BC[22].

The Nasca Monkey glyph, shown below, appears to be perched on a secondary glyph made out of lines with a unique pattern. This portion of the glyph -the gray zigzag line in the picture- is almost a line for line replicate of the SPO glyph. The dotted arrow at the bottom is a glyph line that points to the Tahiti and Moorea islands which are at the end of the SPO.

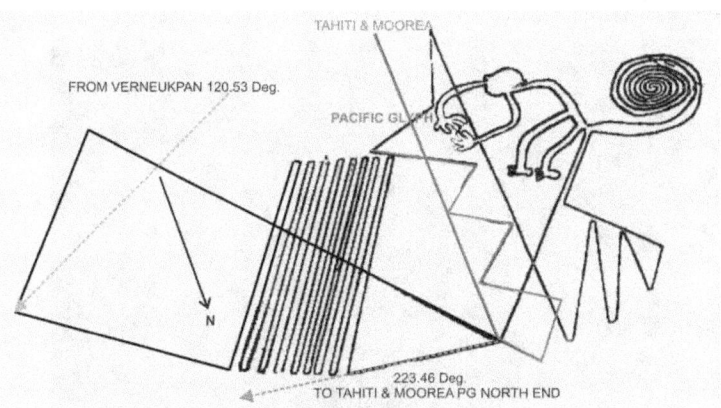

The southern Easter Island glyph is a 500x600 mile area covered by a single serpentine line crossed by a triangular shaped figure.

This glyph was also known to the ancients; it is depicted graphically as part of the Monkey glyph as well. See graphic p.195. The array of parallel lines on the Monkey glyph count differs by one from the ocean glyph. In the graphic on this page, we have superimposed the Monkey glyph onto the map of the Pacific Ocean and Easter Island Glyphs to show the nearly one-to-one relationship that exists between the Monkey and the ocean glyphs.

These graphics are drawn to scale. The Pacific Ocean glyph is shown in gray, the Easter Island glyph in white.

The parallel lines on the Monkey and on the Easter Island Glyph are crossed by a 'triangular' shape. The length ratios of the sides of these 'triangles' for the Monkey and Easter Island glyphs are closely simlar: Easter Island Glyph 1.0:2.69:2.91, Monkey 1.0:2.35:2.45. The graphic, also, shows a dotted line from Verneukpan. This line is a main line part of the Verneukpan layout that cuts across the inverted 'V' at an angle of 120.53° and reaches the Monkey glyph. The Monkey's tail is the other most prominent spiral on the Nasca plain.

Monkey glyph overlay the South Pacific and Easter Island Glyphs

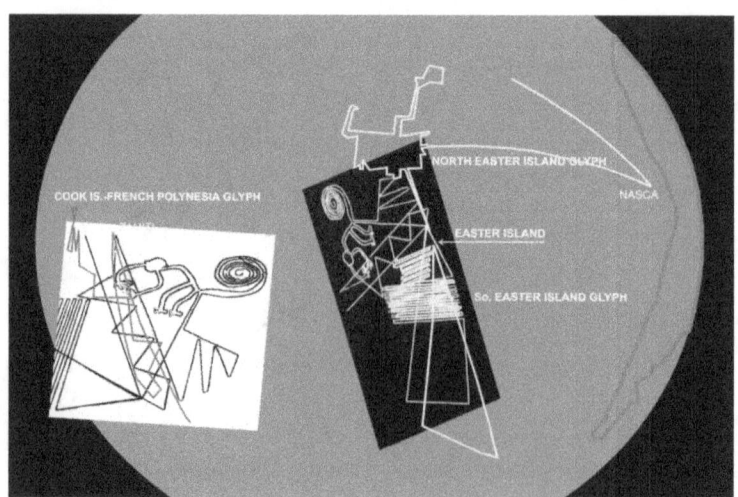

The Pacific Ocean Glyph

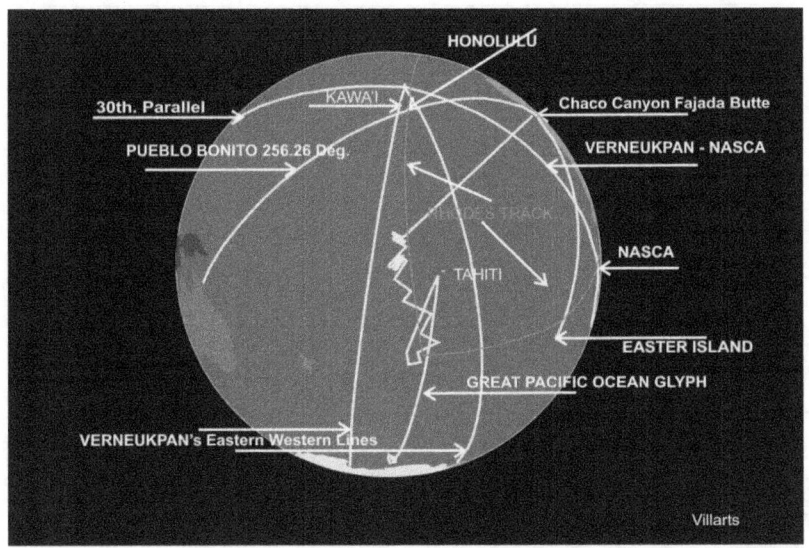

Notes

Chapter 9

AMAZING SITES

TEOTIHUACAN's other connections.

The Teotihuacan Promenade line, on its northward direction, in Wisconsin, US, aligns with the settlement of Aztalan (43.06°N - 88.86°W), and continuing north to Michigan it reaches Copper Harbor and Isle Royale. These two regions, is speculated produced over a billion pounds of copper in 2450 BC[33] which eventually made their way to Europe, perhaps carried by Phoenician Ships.

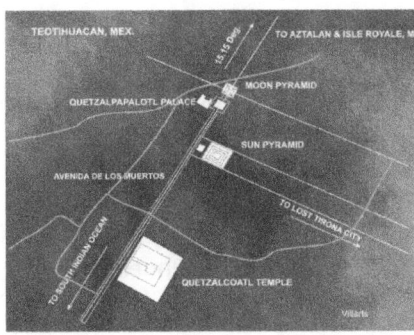

Judging by the petro glyphs found there which depict a vessel with Phoenician type rigging, it appears plausible. It is also reported that the miners moved south to Aztalan during the winter.

The Aztec connection has been previously suggested by various researchers; this alignment strongly supports that theory and as will be explained, the connection is reiterated by several converging facts this research has uncovered. It is interesting to note that the ship glyph was carved using the same single line technique found at Nasca, both, on the animal and the geometric glyphs which were used to depict the ocean glyphs. This establishes a circular relationship between: ocean fearing, copper, glyphs and glyph artistic style. See graphic below.

MYSTERY HILL-AMERICA'S STONEHENGE

The America's Stonehenge site is located south east New Hampshire, near North Salem; it is known locally as Mystery hill. The site has stone structures with echo chambers, sacrificial stone and a central stone that tracks the solstices. This region is believed was explored by the Phoenicians thousands of years BCE. Some artifacts were found in the area, which seem to be of Phoenician origin. (America Unearthed, S1 E3).

The site's encompassing region, when viewed from above at about two thousand feet altitude, appears circular in shape with a diameter of about one half mile.

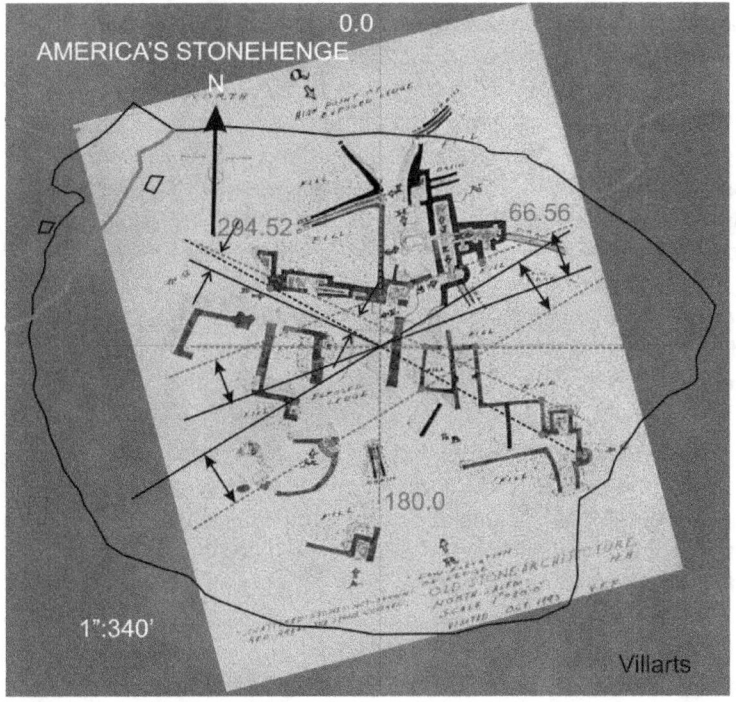

From our perspective the most important feature of the site is that it has paths spreading outwards from its center like spokes on a wheel creating a 'star' which strongly resembles those found at Nasca and the El Infiernito's megalith arrangement, shown on page 109. To create the graphic of the site, shown above, we borrowed a map by V.F. Fagan an Arquitect-1943, and superimposed the pathways that makeup the 'star'. We found that the angles these lines follow match the angles of some of the walls or corners of the structures at the site. We traced the great circles (ellipsis) for these lines around the world and found the location's antipode region; it is in the South Indian Ocean enveloped by the South Indian Ocean Glyph; it is shown in the graphic marked with crossed arrows -p.184. Three of the site's lines are of particular interest. Two run exactly in the direction of the cardinal points, while the third has an azimuth 66.56°. This is the value of the angle of the ecliptic. Drawing a line perpendicular to the 66.56° line at 336.56° corresponds to the declination angle of earth's axis. We recall this arrangement was found at other locations: Nasca, Machu Picchu, Easter Island, El Infirnito and Sacsayhuaman. However, the axis line is not present at the site, so we drew it. Following this line on its northward direction, in Canada, it finds two sites where cairn arrangements are found. One is the Rankin Inlet Inukshuk cairn, which stands alone. Further northwest near Nunavut there is a group of cairns; one is called Innunguaq and another group with nameless Inukshuks. Further north on the line is the gold mine of Medowbank.

The Circle of Illumination at this latitude has an angle of 327.14°. The angle of the sunrise at the summer solstice is perpendicular to it at an angle of 57.14°. This is one of the radii lines marked at the site as a path which runs at about 56.09°; the black line in a north east direction, shown on the graphic on p.201.

Following the 180° line south to Perú it passes fifty miles east of Sacsayhuaman. Continuing south on this line we find the Tintaya copper and gold mines. A few miles west of the mine are the Inca cities of Apacacho and Mauka Llakta (14.92S 71.56W) and the Raqchi Citadel and Temple (14.17S 71.37W). Raqchi and Mauka Llakta's construction style is reminiscent of the Chaco Canyon Pueblos; they have many Kiva like round structures in them. The Chaco Pueblos are discussed on chapter 11. Following the line further south towards the coast of Chile, we find the salt lake Salinas. Following the line on its 360° northward direction it finds the Tetford mines in Ottawa, Canada. Continuing across over the North Pole into China, the line finds the Tongwan ruins and the

Jingbian Burial Mound (37.59°N 108.85°E), also, the Shaanxi Imperial burial pyramids (34.34°N 108.6°W). This type of mound is found in various places in China, they are burial sites for early emperors or their families[26]; they are characterized by their terraced pyramidal shape similar to those found in Egypt and Mesoamerica. South from there, in Viet Nam the line passes forty miles east of the My Son Temples and the Rocks on the Field site.

The Kivas' found in Raqchi are aligned in rows with a bearing of 16.6°, following this line to Viet Nam it also, finds the My Son Temples and the Rocks on the Field archaeological site. The rocks at this site and terraced walls have been compared to those of the Cusco and Machu Picchu regions in Perú. It is significant that these Vietnamese archaeological sites are mapped by these two alignments: the America's Stonehenge-Raqchi 180° - 360° alignment and the Raqchi Kivas' alignment at 16.6°. These alignments are important; the two points at which these great circles intersect, which are generated by the two alignments, are marked with structures of archaeological significance. Raqchi is the site of the Temple of Viracocha. Viracocha is the god associated with Tiamat and Toth and the Aztec serpent Quetzalcoatl, the serpent also a symbol of Chinese mythology. There is an odd number of 'spokes' on the America's Stonehenge 'star' arrangement. The odd spoke runs northwest at 294.52°, a line drawn in this direction goes over the Algonkian petro glyphs in Woodview, Canada. Further northwest in Lake Huron on the shore, the line crosses over other petro glyphs and on Lake Superior finds the copper mining region of Copper Bay and Isle Royale, where the Phoenician ship glyph, shown above is found; this strengthens the circular connection we spoke of before. This copper region is 'targeted' by multiple other sites; the most notable being Teotihuacan, discussed earlier. The west bound 270° line crosses over the Chaco and Chelly canyons and the Blythe geo glyphs in California. We discuss these canyon pueblos in chapter 11. Another line runs northeast at 56.09°, in France finds the covered monoliths of Dampsmesnil and the dolmen of St. Genevive, further down in the Mediterranean Sea in the isle of Rhodes finds Apollo's Temple and across the sea, in Israel; finds The Temple on the Mount, the six thousand year old city of Steph and Bethlehem: Jesus Christ's birthplace. The same line on its SW direction at an angle of 236.09°, in North Carolina crosses over the Judaculla Rock, which will be discussed next. This rock is renowned for its numerous glyphs. Some of the glyphs are quite reminiscent of the glyphs found on the Valle de Azapa Mountain in Chile. A prominent glyph on the rock, found on the bottom center of the rock, is similar to the Chaco Canyon glyph, shown at the beginning of chapter 11. The rock is dated about 3,000 BCE[24] Continuing this line across the Gulf of México, inland it goes over the Toluquilla and the Las Ranas archaeological sites (20.89°N 99.52°W) in Mexico. Following the line with an azimuth of 66.56° on its SW direction at an angle of 246.56°; in Mississippi it passes fifty miles east of the Pharr Mounds and in Louisiana and Poverty Point. In México, the line passes near, the Cerro Blanco petro glyphs and on the Pacific coast line finds the Marismas Nacionales, mentioned earlier.

THE JUDACULLA ROCK
Besides the stone's glyphs, mentioned above, which are believed to be Late Archaic[24] there are several cup indentations of various sizes. Some of these indentation arrangements have been identified as corresponding with the stars of some constellations. One of the proponents of this is Mr. Craig Welch, (umich.edu.) Mr. Welch also equated the Judaculla rock cup glyphs to the Doddington Rock glyphs in Scotland. We took his suggestion further and connected these two locations with a line and continued it westward across México, there we found the line passes by the Teteles Pyramids (18.63°N 97.72°W). This is the third archeological site alignment in México with the Judaculla Rock, we discussed the other two which start at America's Stonehenge. A fourth connection is with the 'Ball Court' at Chichen Itza. A circle following the Ball Court's bearing of 17.39° passes, ten miles from it. Close scrutiny of the 'cup' indentations

on the rock resulted in a discovery: the layout of the larger cup indentations on the rock has the same pattern as the location of some of the major archaeological sites in México. The rock appears to be a map; on the graphic below, note the dotted lines on either side of it and compare them to the Mexican sea shores on the graphic p.206. The dotted lines highlight the lines on the rock. The first photo below is the rock without markings. The second graphic on the next page highlights the map. The graphics are drawn to scale.

Wikimedia Commons. Overlay Graphic by Author

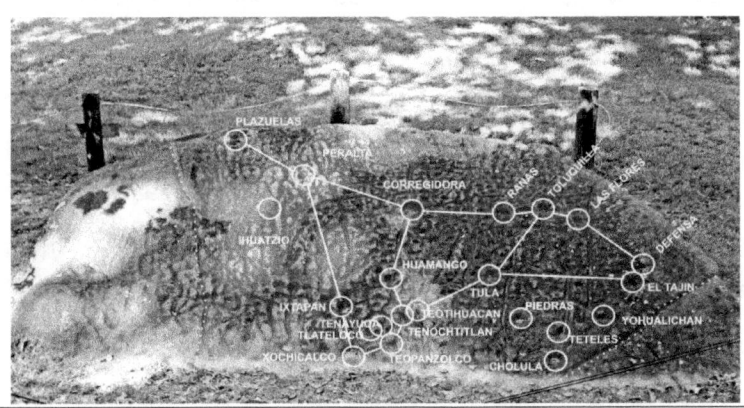

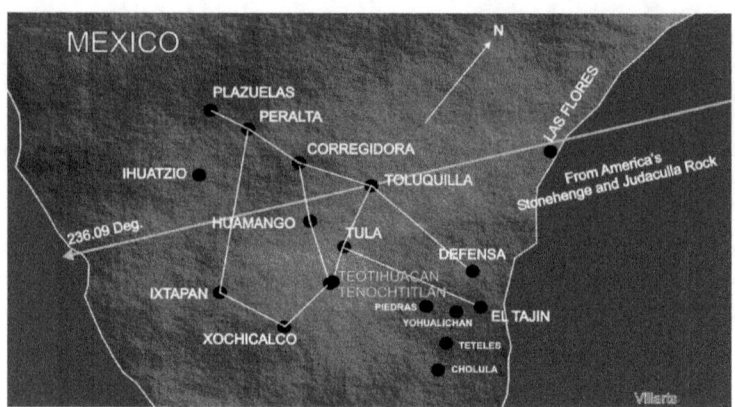

THE AVEBURY CIRCLE

The Avebury circle (51.428N 1.85E) is actually a composite of three megalithic rings. The megaliths are arranged radially like 'stars' with center hubs in a way similar to the El Infiernito and America's Stonehenge. Refer to the graphic on the next page. The Reverend R. Warner, in his book *The Pagan Altar* (1840) argued that both Avebury and Stonehenge were built by Phoenicians[39]. The circle is crisscrossed by two roads; A-436 headed northwest and High St. Green St., headed northeast. In the northeast quadrant near the center there is a megalith we identify as Hub 1 (51.429N 1.854E). Across High St., south of it is a smaller megalith (51.42N1.853E), we named Hub 2. The hub 1 megalith aligns at an angle of 303.23° with the third megalith in the north-west quadrant of the outer stone circle, just west of the Avenue. A great circle drawn from hub 1 at this azimuth aligns with Isle Royale in Michigan near Copper Harbor, the site where the Phoenician ship megalith is found, discussed earlier on chapter 9. Hub 2 aligns with the fourteenth megalith at an angle of 298.16°, a great circle at this azimuth aligns with Aztalan in Wisconsin, US. These facts could add credence to Reverend Warner's claim. On the western side of A-436 at the edge of the henge there is a large megalith starting the circle counterclockwise, we assigned it number one. A line traced from it southward along the avenue across the henge and ending at a megalith outside the henge on the south side has a bearing of 337.16°, near the value of the declination angle of earth's axis. A line nearly perpendicular to the Avenue starting at hub 2 and aligning with two other menhirs in the inner and outer circles, in a westward direction (246.44°) reaches the Moray Inca (13.33°S 72.19°W) in Perú at the center of the larger amphitheater shaped structure crossing it lengthwise. It may be recalled this structure is on the Easter Island - Cheops' Pyramid line. The layout of the Moray structure is fashioned after a Lemniscates curve. The 90° direction is marked starting at Hub 1 connecting two shorter amorphic megaliths; one a few feet from the hub the other 182 feet away to the east. The Hub 1 megalith connects with another smaller megalith located also on the east side, at the edge of the circle on the NE quadrant about 200 ft. north of Green St., at an angle of 66.16°; nearly equal to the angle of the ecliptic. This line on its westerly direction 246.16°, in Perú, finds the Sacsayhuaman's Sun Dial at its center! It may be recalled the Sun Dial also depicts the declination angle ofthe earth's axis. Another line connecting a small stone on the outer circle twent feet from Green St. with hub1 has an azimuth of 265.0°. A circle drawn at this azimuth reaches Easter Island at Ahu Ko Te Riku,

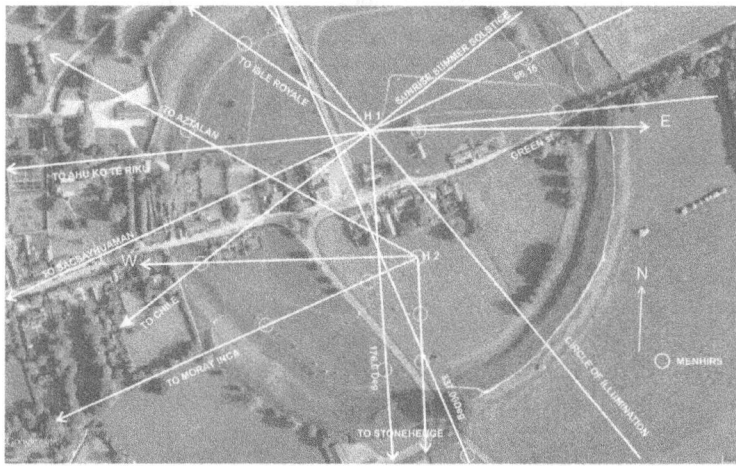

matching the angle of the Ahu on the eastern side. Avebury shares similar geometry found at other locations; a radial arrangement that displays the obliquity angle of the earth's axis and the cardinal points. At Hub2 the Cardinal points are also marked, with a smaller stone inside the southeast quadrant and another to the west on the outer ring which is a megalith shared with Hub1. This megalith is at about 230.6° from Hub1 and 270° from Hub2. At Avebury the *Circle of Illumination* has an angle of 320.34°, consecuently, during the summer solstice, at sunrise the sun's theoretical azimuth has an angle of 50.66°. At sunrise on June 21, 2014 the sun was at an angle of 50.7°. The sun's direction line at the solstice continued SW has an angle of 230.6°, therefore the sunrise at the solstice is marked by hub 1 and the megalith shared with hub2 on the outer circle. The same line in this direction, in southern Chile passes 59 miles northwest of Monteverde Creek the site of 14,800 year old human remains.

Neither of the Hubs is located at the center of the circle, however, the radii emanating from Hub2 share the larger megaliths of the outer circle or the markers where they once stood. Both hubs align with Stonehenge. Hub 1 aligns, after crossing over the White Horse geo glyph about four miles south of Avebury, with Trilithon 25 at an angle of 176.01° and with megaliths 45, 80 at the center and 12. Hub 2 aligns with a line at an angle of 176.09° the Stonehenge alignments are shown in the graphic on page 207. The alignments at Avebury are shown in the graphic above.

STONEHENGE

Stonehenge has been analyzed in many ways. In Chapter 6 we introduced a geometric analysis which showed Stonhenge is part of a global arrangement which forms a global clock. The discovery was based on the analysis of topographic measurements of the orientation of its megaliths with respect to the sun's seasonal positioning. We also showed that the alignment of some of its megaliths correspond with the angle of the Circle of illumination at its geographical location. Here we provide further analysis, as we have done for every site in this study. In the graphic below we show some of the geometrical megalith arrangements in relation to other sites. Its alignments are many; here we will discuss some of the most important alignments that were found. We superimposed the alignments over a graphic by Anthony Johnson[39], and use his nomenclature for the analysis. The sun, at sunrise during the summer solstice in Stonehenge, aligns with SW Trilithon 56 and Heel Stone 80 with a bearing of about 50.61°(230.61°), as was reported in chapter 6.

Basic Graphic by Anthony Johnson. Wikimedia-Overlay Graphic by Author

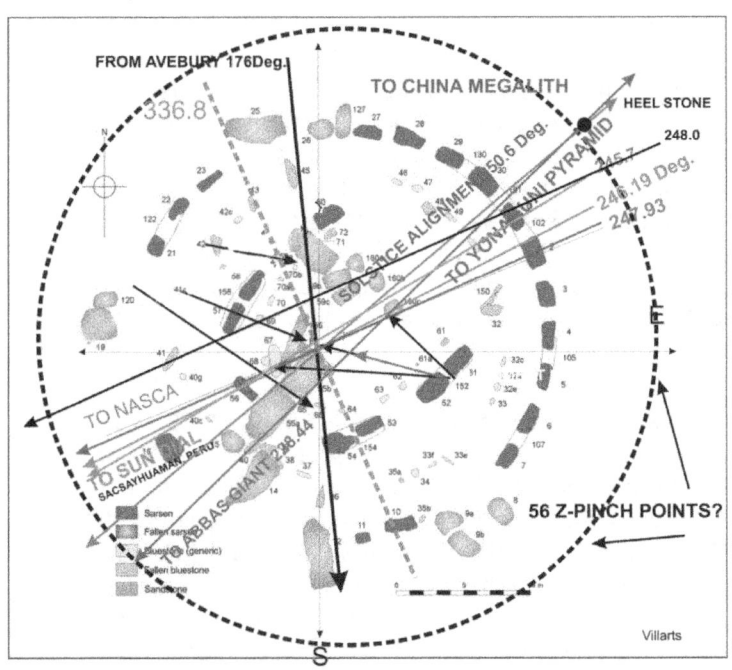

Drawing a circle with this azimuth in the northeast direction it passes by the Fosbury Hill Fort, continuing over the North Pole, now in a SE direction over China the line finds the Phallic Megalith near Ningxia. In Japan the line passes about one hundred and sixty miles south of the Yonaguni underwater pyramid (24.4°N 123.29°E). A circle drawn from the Heel Stone through Trilithon 101 and parallel to Trilithon 152, in this heading, southwest of Stonehenge aligns with the Abbas Giant at an angle of 228.44°. Following the circle in the opposite direction at an angle of about 48.44° other alignments are made. Starting at the Abbas Giant with the Trilithons 101and 152 and the Heel Stone in the UK, continuing to Japan the circle now aligns with the Yonaguni Pyramid. A line going through the gap of Trilithon 102 at an angle of 246.19°, aligns with the Sacsayhuaman's Sun Dial in Perú. A second line with a bearing of 246.46° aligns with the Moray Inca. The value of this angle is about equal to the earth's ecliptic. In the graphic we also find megaliths aligned at the angle of declination of the earth's axis: megaliths 25, 43, 64, through Trilithon 154 (53&54) and 10 on the SE. Megaliths 2, 80, 15 & 16, align at an angle of 240°, a great circle in this direction when followed to South America, crosses the Sajama desert. This alignment is reciprocal, at Sajama the line aligns with a line forming node 107 and connects through nodes 108 and 110. The importance of alignments through the Sajama desert is discussed in an upcoming publication.

Stonehenge's line on the Nasca plain located at 14.69°S 75.12°W is found by following a third line headed south west at an angle of 247.93°, shown in the graphic, p.209 which aligns with Thriliton 2, megalith 160c and 80. Another line connecting megaliths 1, 59b, 69, 57 and 41 has an azimuth of 248°. A great circle in this direction, reaches the Nasca Calendar Wheel. In the graphic, the cardinal points are defined by megalith alignments; North South is defined by the alignment of megaliths 26, 80 at the center and 12. The East West alignment is defined through Trilithon 105, 80 at the center and 19.

Surrounding the megaliths there are fifty six pot holes where it is believed poles once stood; however, some scientists believe that a cosmic event of extremely high energy, which can be demonstrated in the laboratory, produces an energy discharge with peculiar chalice shape, around a sphere known as the 'z-pinch'. The energy discharge produces energy 'tubes' that burn a circular pattern on a surface. The number of 'tubes' or burnt spots is usually fifty six. This fifty-six spot circular arrangement is found in glyphs in various places around the world; the Nasca Calendar Wheel being another one of them.[27] They believe that this event may have occurred and observed in antiquity, which prompted the observers to record the event in monuments and glyphs.

THE ABBAS GIANT

The Abbas Giant is a large 'caveman' glyph, about half the size of the Candelabra and like the candelabra the figure is painted on the side of a hill. It is 237 feet from his toe to the top of his mace, it is located near Cerne Abbas in southern UK, south west of Avebury at (50.8°N 2.47°E). Like the candelabra it is also a guide post to important places. The most prominent feature of this glyph is his mace. To the east of the glyph there is a quadrangle whose western line has a bearing of 336.0°, close to the value of the declination angle earth's axis. A line nearly perpendicular to this line with a bearing of 247.43°, on its westerly direction, on the Nasca plain finds the glyph's line which ends at the same prominent node the Nasca earth's axis line 336.84° starts. Following the line in the opposite direction NE 67.43°, in Tibet we find the region near the Great Potala where T. Lobsang Rampa, in his book "The Third Eye", describes the cave of the Giants is located. In Thailand passes near the temple Prasat Ban Ben, in Indonesia the circle passes forty miles west of the Niah Cave and in Australia fifty miles west of Lake Mungo, a UNESCO man origins site. These last two locations, mentioned before, are where human remains older than 6,000 years old were found.

Wikimedia Photo by Pete Harlow. Overlay Graphic by the Author.

The giants right thigh distal line has a bearing of 90°. An ellipse drawn in this direction, in Crimea passes eleven miles north of the 2,500 year old city of Chersonesos built by the Greeks. In India, finds the temples of: Rukshamaniji, Somnath, Tambdi Surla, Cheluvarayaswami, Narashimha, Srirangathanswami, Mongeswarar, Sivancovil and Thirukedishwaran. In Sri Lanka crosses the Polonnaruwa temple complex with the reclined Buddha and a grand Stupa and ten miles west of the Budhdhangala Temple and Stupa, twelve sites in all, align with the 90°circle. The continuation of the line in the 270° direction cuts through Costa Rica near the Turrialba Valley where a civilization flourished dating back to 10,000 -7000BCE.

The mace on the Giant's right hand points south east at 116.92°. Following a great circle with this bearing, in France passes fourteen miles north of the covered monoliths of Dampsmesnil and seven miles south of the Dolmen of St. Genevive. These two sites have been previously triangulated by other alignments. In Greece the circle crosses the amphitheater at Stratos dated fourth century BC and in Crete goes over the Palace at Knossos. Across the Mediterranean it continues through Alexandria and over the Great Pyramid of Giza nearly parallel to its diagonal and passing at the feet of the Sphinx. This line in the South Indian Ocean follows the SIO Glyph parallel to its central section, for 1,375 miles. The giant's phallus connects to the other important phallic symbols (megaliths) in China and Colombia, we had discussed earlier. The great circle connecting with the Chinese phallic megalith passes two miles south of Stonehenge and four miles north of the Figsbury Ring and near the Bulford Stone and the Fosbury iron age fort. The line which connects the Abbas Giant with the El Infiernito phallic megalith, on its easterly direction at an angle of 78.51° finds the flint mines in Grimes. In India finds the Khaurhajo, Neelkanth and Lingaraj Temples region, renowned for their sexually explicit statuary and continuing southeast in Sumatra passes ninety miles from Aur Duri. Continuing onto southern Chile the line passes near Monteverde Creek, which also was previously triangulated, further north crosses the salt lakes region in Cordoba Argentina. Following the line in the opposite direction, in Cornwall near Minions finds the Long Tom phallic stone.

ATACAMA GIANT

The Atacama Giant (19.948°S 69.633°W), as the Abbas Giant, is also carved on the side of a hill. This one is an outcrop in the Atacama Desert south of Azapa called Cerro Unitas in Chile. The glyph is about 276 feet tall, although some references describe it as being over 300ft. The giant's sagital axis line has a bearing of about 62.89°. At this azimuth it points to the Temple on the Mount in Jerusalem. A great circle drawn in the direction of the right torso line continued over the right side vertical antenna reaches the Pyramid of Cheops. Following the right leg medial line at 57.43° to Turkey it reaches Göbekli Tepe. A line closely following the angle of the left arm has an azimuth of 90.0° in this direction, seven miles east finds the archaeological region of Tarapaca covering approximately 70 square miles. The region contains probably hundreds of glyphs, stone circles, squares, animal glyphs and the ruins of ancient settlements. A line drawn from his left knee through the skirt line intersection with the torso line reaches the Abbas Giant in a north east direction of 35.64°.

The skirt line runs at 337.87° a circle on this northwest direction cuts through the Azapa valley famous for its numerous geo glyphs and petro glyphs, in particular, those on the "Cerro Sagrado" The Sacred Mount". Continuing north the line passes by Machu Picchuand in the US, crosses near Poverty Point Louisiana. In China the line passes near the Astronaut Rock of Silun near Guangdong. The figure's left torso line cuts across northern Libya. Near the border with Egypt and the Mediterranean Sea the line cuts through a desert area where anthrophomorfic glyphs outlined with stones are found. One is possibly a Tyranosaurus Rex (500ft.) and another, a Terodactylus. A short distance away (31.652°N 25.084°W) there are a series of ground scraped lines in an arrangement virtually identical to lines south of the Atacama Giant on the same hill (-19.951°S 69.633°W).The Atacama Giant is shown below, in graphic over Google©earth map.

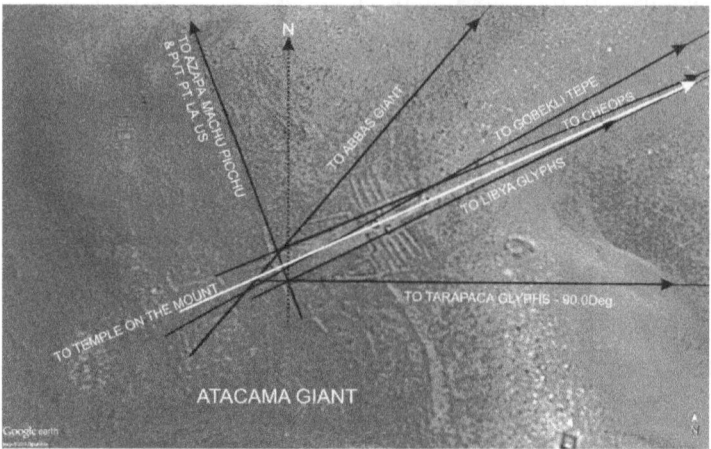

ATACAMA GIANT

THE BLYTHE INTALGIOS

These geoglyphs are located in southern California, US at 33.794N° 114.536W°. The Department of Interior describes them thus: "There are a total of six distinct figures in three locations, including a human figure at each location and an animal figure at two locations. The largest human figure measures 171 feet from head to toe. Geoglyphs are difficult to date, so archaeologists have no way of knowing their age".

The graphic below shows the site where the 'horse'- The four legged figure could be any animal, but a horse is likely*- appears with a double spiral, the site shown is NE of the other two; the site NW of it has a similar man but no 'horse', the other site found SW of it has a 'horse' but no spiral.

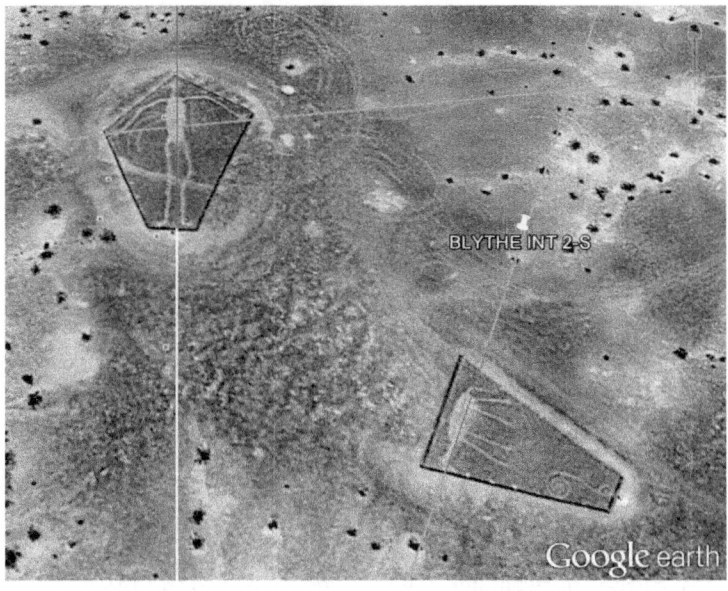

* "The last prehistoric North American horses died out between 13,000 and 11,000 years ago, at the end of the Pleistocene, but by then EQUUS had spread to Asia, Europe, and Africa". Jay F. Kirkpatrick and Patricia M. Fazio, July 24, 2008 in Livescience.com

For the geometric analysis we found the bearing of the man's position in a manner similar to what we used for the Abbas and Atacama Giants. The largest Intalgio, as these glyphs are commonly referred to -not shown- is about the size of the Candelabra glyph whereas the other giant glyphss mentioned earlier are about fifty feet larger. The three men figures are arranged in a triangular fashion almost forming a right isosceles triangle with one of its sides pointing east at 86.77°, the other at 347.45° and the hypotenuse at 36.29°. Of the three men glyphs only one (man2) has its sagittal plain pointing north at 360°; the other two; one points at 206°(man3) and the other at 163°(man1). Men one and two are accompanied by 'horses'. Man2's horse has a spiral at his feet; the horse's body is angled SW at 205° and the spiral is perpendicular to it at 115°, while man1's horse faces SW at 220°. We traced great circle ellipses for each one of the directions given above to find alignments.

The man's sagittal plain facing north points out that all the figure's directional alignments were not casual. The three men have their arms fully extended, but all are a few degrees off being perpendicular to their sagittal axes.

Starting with the triangle's side in the NW direction at 347.25°, 167.25° the following alignments were encountered: In the Baja peninsula México the petroglyphs of San Francisco and San Pablo. In India the temples and burial sites of: Badami Shri Dattatreya, a large cave and temple complex comprising frescoes and massive carved columns out of the rock similar to the city of Petra and the Stupa of Stadhara, a site which comprises fourteen Stupa structures, further north in line are the Taj Mahal and the temple of Trakeshwar Mahadev. Back in the US in the State of Oregon, the line goes over Kennewick the site where a 9,000 year old man of Polynesian extract was found. The triangle's east-west line, continued in a circle in the easterly direction at an angle of 86.77°crosses over Poverty Point in Louisiana, US and in South Africa finds the caves at Sterkfontain. The third side of the triangle describes a circle at an angle of 36.29° which in this direction, in Ireland finds the Dolmen at Lough Scur, the Cairn and Megalithic Tomb of Loughcrew and the Tara Stone. In the UK cuts through Avebury, then, passes near the Wiltshire Fort and in France passes fourteen miles south of the Dolmen of St. Genevive. Crossing the Mediterranean Sea, in Libya the circle crosses atop the Temple of Zeus at Cyrene and thirty miles west of the Temple of the Oracle at Siwa and in Sudan finds the Pyramids of Mèroé.

Man1's Sagittal Plain points NW at 343°, following this direction in Paryangthe in the Himalayan Mountains finds Mani Stones - Buddhist prayer stones. Further south in India it finds the temples of; Hrishabhdev, Virat Mandir, Ghampur and Kanjamalai Pernumal. An ellipse in the direction of the man's Coronal Plain points 71° NE; in this direction in Arizona finds the Montezuma Castle Dwellings, the Newspaper Rock and the Puerco Pueblo. In New Mexico crosses over the Chacoan Casmero Pueblo the Bandelier Cliff dwellings, the Alcove House and the petroglyphs at Tsankawi, in Illinois the Modoc Rock Shelter dated 9000 years, in Indiana finds the Angel Mounds and in Kentuky Mount Horeb and in South Carolina the Charleston Mound. Around the world in Australia encounters the petroglyphs at Nepabunna.

Man1's horse appears to be heading in a direction of 220°; following the great circle in the opposite direction the line goes over the Grand Canyon's North Rim, in Utah it finds Anasazi Roadside ruins, in North Dakota passes by the Standing Rock Mounds and on the border with Canada the Manitou Mounds. Across the Atlantic, in south west France the circle goes over the Sacrificial Table at Finistère, further east encounters the Carnac region the site of many menhir alignments, continuing through

France it goes over the caves of Lascaux and twelve miles from La Chapelle Aux-Saints. Crossing into the Mediterranean Sea in the island of Sardinia in the town of Lunamatrona, the circle goes over the stone ring of Trobas and the Tomb of the Giant at Nixias. This site is of special significance in that its layout is similar to the Blythe Man' body: a double stone row ending on a head stone topped with an arched stone row. The Tomb in Lunamatrona is aligned with the summer solstice; the sun illuminates the chamber at 1245hrs on June 21. In Hawaii the ellipse crosses over Lana'i's Pu'upehe ritual Platform.

Notes

CHAPTER 10

The Pyramids of Giza & the Earth's Plate Tectonics

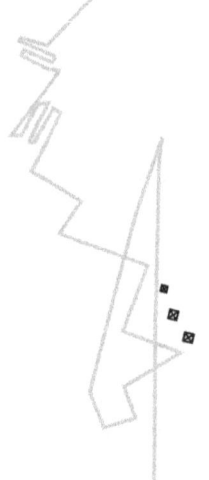

In chapter five we discussed some of the alignments the pyramids of Giza have, specifically, the alignment of the pyramid of Cheops with Ahu Akahanga and the Abbas Giant. The apex of this pyramid is the point where their great circles (ellipses) intersect: The circle drawn which follows the direction of Ahu Akahanga, in Easter Island (EI) and another which follows the direction of the Abbas Giant's mace, as discussed in the previous chapter.

This geometry, we argue, fixes the relationship between the locations of these three archaeological sites; which lead us to conclude, these sites are related to each other in a planned design. The Ahu Akahanga circle, as reported, could be attributed with the purpose of setting the latitude for the location of the pyramid of Cheops.

Earlier in the study we showed that the arc segment of the Ahu Akahanga ellipse which connects the pyramid of Cheops with Basarah, reaches its northern most latitude of 30.4°N at a point nearly half the distance between the two locations at longitude 41°E. The pyramid of Cheops has latitude 29.975°N and longitude 31.140°E. The Abbas Giant's mace's circle centers the pyramid ten degrees west of the maximum latitude. The immediate question this arrangement prompts is; what determined the geometrical position these two great ellipses have? One can argue that Easter Island being relatively small and at a distance of about 16,000km. from Giza, could be considered to be a point and any ellipse passing through it could be jogged to match the pyramid's apex. However, the Ahu marks not only the specific point in the island the arc should pass through, but also its direction. Therefore we conclude this anchor point was designed and built with this purpose. In this work we have reviewed several other alignments of similar importance and purpose; with other Ahus in the Island and several structures elsewhere.

The case for the Abbas Giant circle is different. It can be argued the Giant could have been drawn anywhere pointing his mace in any direction, if it not were for the fact that the arc the mace traces is a segment of a great circle which follows almost 1,400 miles (~2300km) of the South Indian Ocean ridge line (SIO Glyph). The two facts just described lead us to conclude that the pyramid's location was very likely selected as prescribed by the ridge line and EI's Ahu Akahanga, and venture to say that the pyramids' arrangement and north alignment are probably not due to stellar alignments, as has been claimed for many years by a number of researchers and writers[13].

However, the foregoing explanation of the geometry involved in fixing the site for the pyramids said nothing regarding the physical layout of the site or the number of structures present or their shape. The stellar alignment with the constellations of either Orion or Cygnus, the various researchers claim they have, remains unresolved and in doubt nonetheless. We sought to find a rationale that would explain the diagonal alignment of the pyramids that makes the alignment with Baalbek and Mount Ararat possible, as had been claimed[2].

The follow-up study resulted in further discoveries and data that reaffirms the original argument and explains the physical layout of the complex: the pyramids' location, heights, bearing and distance between them. These variables are dependent on a different set of earth's trench or ridge lines in the ocean, they are located in the South Pacific Ocean at -25.250714° -154.263026°, the ridge lines can be seen with Google© earth by viewing the ocean closely without historical mapping or by downloading a file[30] which overlays the ocean lines.

The first question we sought to answer was the reason for the north-south alignment for the three pyramids. That alignment as well as their claimed alignment with constellations exhibited the same problem; the North Star Polaris is as much a moving target as the constellations are; which we argue diminishes the possibility of an ancient civilization having designed a layout of this magnitude based on alignments with stars -moving targets, particularly with the presumed limited technology available to them, despite the myriad theories that exist of how it was done. One such theory quoted in Building the Great Pyramid[53] is Otto Neugebauer's "On the Orientation of Pyramids" In his theory he proposes a wooden or stone model called a pyramidion; "The pyramidion is precisely aligned north to south" and the shadows cast by it by the sun at various times." The method says nothing about the time element involved over the twenty to thirty years[9] it is estimated it took to build them with crews of up to 4,000 men[10]. Again, the wooden model relies on Polaris and as mentioned there the method was only useful in winter.

Creating a geo-centric referencing system is the subject of study by the International Terrestrial Reference System, for which various datum have been developed to create global standards for global positioning, aeronautics, space, defense, geographic and point of interest POI applications-. "The International Terrestrial Reference System (ITRS) describes procedures for creating reference frames suitable for use with measurements on or near the Earth's surface. This is done in much the same way that a physical standard might be described as a set of procedures for creating a *realization* of that standard." Wikipedia.com

From the data collected earlier, it became apparent that the positioning of the tectonic plates was used for this purpose in ancient times. The current work appears to confirm that belief and strongly suggests the ridge lines could become a useful standard like other geodetic standards the ITRS uses, such as triangulation pillars or bronze disc markers placed in various areas of the earth indicating coordinate points accurate to ten decimal places.

In the first part of the research, we established a procedure to initiate the geographic evaluation of the positioning for any structure within a site; first we determined if the structure is aligned with the Cardinal axis and which feature(s) of the structure are involved in the alignment. The Giza pyramids' alignment with the Cardinal points is common knowledge, well documented by numerous studies found in the literature, the reference given is a fairly comprehensive review of pyramid facts, many such references are available each reflecting its preferred theories.

In the current study we used the Chephren pyramid to trace global ellipses following the Cardinal points, each ellipse starting at the southeast corner of the pyramid, at 90° from each other; thus their intersection point, half way around the world, determines the pyramid's antipode point on earth.

In the chapter Ocean Lines, we described the ocean line formed by the earth's ridge line south of the French Polynesian islands of Moorea and Tahiti in the South Pacific Ocean (dubbed SPO Glyph)

—a tracing of it is shown above, p.221. In the chapter we discussed how the ridge line's shape had been copied and engraved as a glyph at Nasca by the ancient authors of the Nasca Lines: it is the perch on which the famous Monkey Glyph stands. The graphic shows that the ridge line is composed of a vertical line which is located at the 150° longitude meridian, which starts in Antarctica and ends, just south of Tahiti. There it folds back going south-west, then returns northward in a zig-zag manner. The pyramids' antipode point is located at the second fold of the ridge line at about 30.0°S 149.0°W. In the graphic we also show the location of this point with respect to the ridge line glyph. In the graphic we enlarged (~650 times) a graphic of the pyramids drawn to scale to depict their relationship. The ocean ridge line has two important characteristics: The vertical line runs north-south, and is located exactly on the 150° longitude meridian. The pyramid of Cheops location is at about 31°E, so, at its antipode point its longitude is about 149° and the pyramids' NS alignment runs parallel to the ridge line. This line of the SPO Glyph is what we believe determined the North alignment of the pyramids and every other structure thus aligned; such as the Temple of Kalsasaya in Tihuanaco, instead of their alignments being based on the mobile magnetic pole or the Polar Star. The SPO Glyph has a natural earth line that happens to precisely run north south; as such it is a perfect 'immutable' anchor, which at this point of our research appears to be unique amongst all ridge lines; if anything it is perhaps the longest and most obvious line. We believe this line, which runs north all the way from Antartica for 6,000km, is the Ancient's original geo-centric based source for the "North" concept.

We were able to safely come to the conclusion that the Pyramids' geometry is related to this Glyph by surveying the other components of the ridge line; the other features of the ridge clearly show the connection the pyramids' layout has with the entire ridge line's layout, i.e. the nearly one to one (1:1) relationship that exist between all their geometric measurements; geographic alignment and positioning on earth, the heights of the pyramids with the glyph's peaks' heights, and the distance between them. In addition, the overall layout's pattern is similar which is shown below and as a white 'step' line in graphic 1, p.227.

Now the comparison in numbers: The line that follows the 150° longitude meridian ends just south of Tahiti, there as mentioned, it folds back SW at an angle of 195° and continues for about 2,000km. From there it returns north-west following a symmetrical zig zag path, eventually reaching past the island of Moorea. The zig zag line has three major saw tooth-like folds at almost right angles with each other; *the peaks that point east*, not their corresponding troughs. The peaks of the three teeth are distanced from each other as follow: 752km.(1.0) between the largest in the south and the middle one to the north and 716km. (.95) between the middle one and the third to the north of it. The ratio of the distances is shown in parenthesis. The first two peaks are near equally sized while the third one is the smallest. Their heights are: 479(1), 459.4(.96), 285(.59) km. The ratio of their heights is shown in parenthesis. Similar measurements were taken for the Giza pyramids, the distance between their apices are: Cheops-Chephren-490m, Chephren-Mycerinus-440m and the ratio of their heights is (.90). Likewise the pyramids' heights (original) and their ratios are: 147(1), 143(.97), 109(.74)m; comparing the ratios; heights and distance show near correlation with the ridge line's geometry.

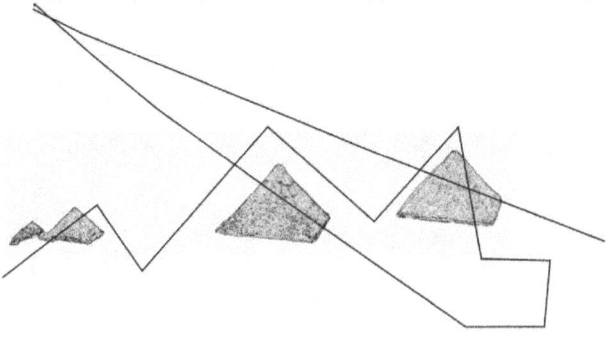

Recall that the pyramids' geographic location is antipode to the SPO Glyph. These numerical relationships can be better appreciated visually in the graphic above and Graphic 1 shown below. In the graphic, superimposed on a map to scale of the pyramids, is a tracing of the earth's SPO glyph also drawn to scale and then reduced close to the pyramids' scale. The graphic of the SPO glyph was moved north along its NS axis, over the North Pole until it reached Giza, then, it was moved about a degree east to align with the Chephren pyramid. It is important to emphasize that at this location the SPO glyph's north becomes its south; as such it makes patent the geometrical relationship it has with the pyramids' layout; the glyph was geometrically translated. In this position the angle of the peaks is nearly the same as that of the pyramids' arrangement ~39°; also each peak coincides with each pyramid in the same order. If the Cheops pyramid were to be enlarged a thousand times it would fit snuggly in the square section of the glyph on the upper left. In the graphic this relationship is shown in dotted lines. This fact addresses the question; why are they square not any other shape? Also, the size of the pyramid's base may have been set to be a multiple of the square section's size; once the shape and size of the base and the height are given all the other measures are derived from these[53].

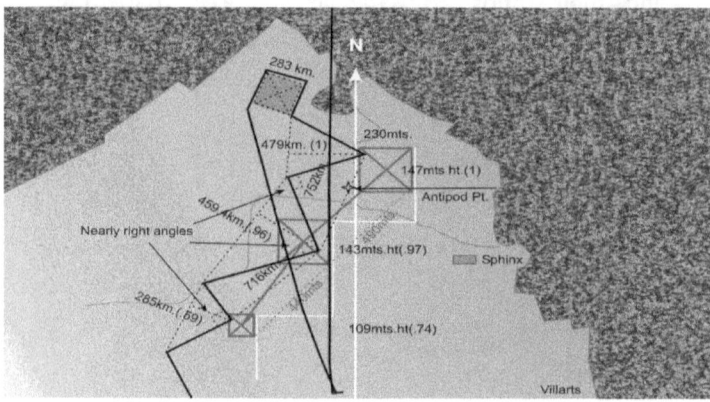

Graphic 1

After reducing the SPO glyph's size a thousand times, curiously, the difference in areas of the squares divided by a thousand and then the cube root taken the result is equal to the number of pyramids:

$$(283^2 - 230^2)/1,000 = 3^3$$

In Graphic 2, the arc segments for both the Abbas Giant's Mace and Ahu Akahanga's are shown as they cross over the pyramid of Cheops' apex. The Abbas Giant ellipse line is about 15° off the pyramid's diagonal. Following is a second geometric analysis.

We took the SPO Glyph tracing and rotated it along its NS axis to obtain a mirror image, followed by a rotation of about 75° clockwise about its center. This transformation yielded extraordinary results. While maintaining the overall alignment of the SPO glyph's peaks with the pyramids, some of its lines became aligned with other features of their layout, this time pointing out their relationship with the Sphinx and the causeway that connects it with the pyramid of Chephren.

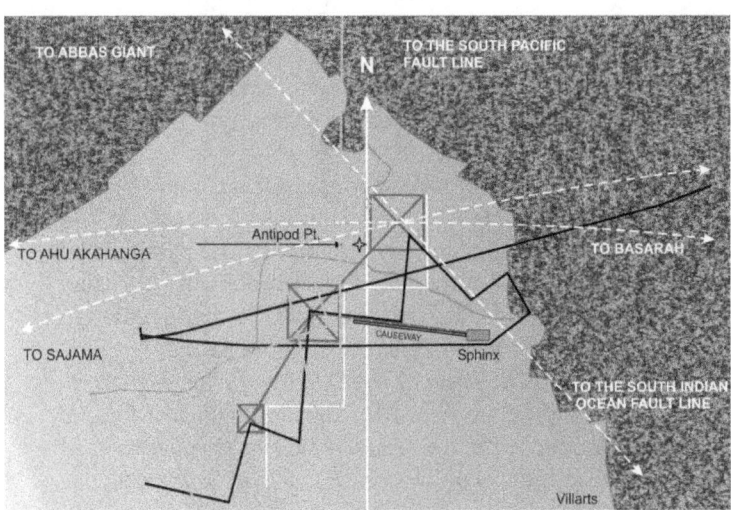

Graphic 2

Simultaneously, it aligned the southern line of the glyph's peak, which corresponds to the pyramid of Cheops, with the diagonal line across the pyramid; which is also closely followed by the Abbas Giant's Mace circle which, in turn, lines up with the SIO Glyph.

This exercise showed the closed reference matrix that exists between all these elements. The geometric layout of the Giza complex reflects the exact location and layout of two of earth's most salient geological features: The South Pacific and South Indian Ocean's ridge lines.

Summarizing; the SPO Gyph sets the 150° meridian great circle that determines the geocentric marker for north; the alignment the pyramids were given, but does not determine their location along its circumference. The Abbas Giant-SIO Glyph alignment circle would have placed the pyramids 8km southeast of Alexandria the point of its intersection with the 30°(150°) meridian. The actual location is approximately one degree south and one degree east from the circles' intersection. This location at the antipode point corresponds to approximately the midpoint of the line that connects the two large peaks on the SPO glyph; which may have determined the one degree displacement in both directions. Ahu Akahanga's circle sets the location of the pyramid of Cheops along the Abbas circle.

The Abbas Giant and the Ahu Akahanga appear to have been designed as geodetic markers to help unravel the logic in the design and layout as was exposed here. Analyzing the pyramids by themselves would have not revealed the underlying (literally) rationale for their location, layout and sizes. The triangulation offered by the two geodetic markers provided the path.

In our analysis, a graphic to scale of the SPO Glyph was aligned NS inverted from its natural alignment in the South Pacific Ocean. Also, it was positioned over the eastern side of the pyramid of Chephren; the alignment line we used to find the pyramids antipode point- shown in the graphic as a star near the southwest corner of the pyramid of Cheops. The geometric translation of the SPO ridge lines results in a nearly perfect superimposition of its layout over the pyramids' site; this indicates that the variable layout of the pyramids (arbitrarily open to the designers to choose; following mobile stars, as is claimed) was instead, fashioned after the immutable ridge lines' layout. The geometric congruency of design between the SPO Glyph and the Giza layout is confirmed by the geometric ability to rotate and flip the glyph about its center while maintaining the alignment with the three pyramids and the addition of two other important alignments; with the Sphinx and Causeway. The pyramids' north alignment is confirmed by the new alignment with the pyramid's diagonal. This new alignment also reveals the Cheops pyramid's lateral pitch; it is found encoded in the glyph as well. The graphic below clarifies this last point as follow:

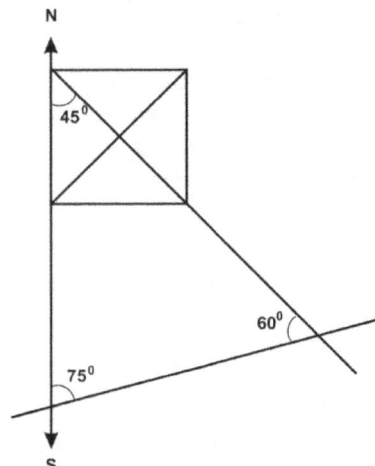

The 'north' line of the glyph was rotated 75° about its middle peak which caused one of the glyph's lines which form the Cheops peak to coincide with the pyramid's diagonal at 45°. Therefore, the angle of the Cheops peak line on the glyph forms a 60° angle with the glyph's north line. This angle is equal to the average pitch of the Cheops pyramid's side[53].

The relationships described are only a small part of a greater globe spanning arrangement.

Graphic 4 below, p.233 is a Google© earth globe map of the SOP and SIO ridge lines, it shows their locations relative to Australia, Antarctica, Easter Island and South America. Superimposed over the map are the alignment arc segments mentioned previously.

The map shows another great circle (labeled Sajama line) which connects the pyramid of Cheops with the SPO ridge line going through the Sajama desert. This great circle will be discussed below. The lower left of the graphic shows the 1400 mile (~2,200km.) segment of the SIO ridge line that is closely followed by the Abbas Giant's circle. The Ahu Akahanga line appears in gray connecting Easter Island, shown (not to scale) in the upper right hand quadrant.

The SIO glyph has a spur line headed southward at 202.79° starting at 42.434°S 89.837°W and runs for 634km. Following this line over Antarctica it crosses the Sajama desert and Isla del Sol in Lake Titicaca -in the ocean Lines chapter we discussed this alignment in detail- the line is shown at the bottom of Graphic 4. The Sajama desert region has been claimed to have been the seat of an earlier civilization[19], specifically it is claimed to have been a settlement of the people of Atlantis.   It is interesting to note that the extension of the SIO spur line intersects the circle that connects the pyramid of Cheops with Sajama and the SPO glyph, at the Sajama desert. That triangulation fixes the Sajama desert as a POI. In an upcoming publication we will cover other results of our research on the Sajama Desert. One of the results shows how features in the desert fix the position of the Cheops-Sajama-SPO circle, i.e. that circle is generated by features in the desert which when extended in a great circle reach the pyramid of Cheops.

Keeping in mind that the Sajama region encompasses the pre-Inca civilization region of Lake Titicaca, Tihuanaco and Puma Punku, it is possible that the ancient structures in this region were also localized following the same geographical scheme as we have discussed was used for the Giza pyramids, perhaps to a lower level of geometric sophistication. In graphic 3 below, the step line (light gray) is the outline of the Akapana Pyramid in Tihuanaco, enlarged to scale, then laid over the Giza layout.

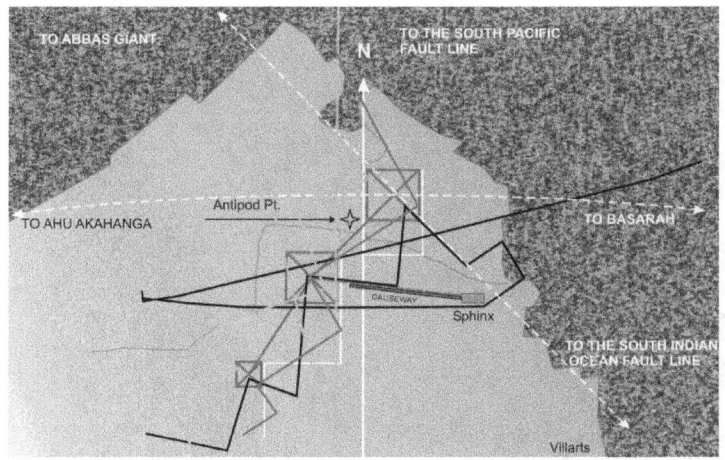

Graphic 3

Although the peak ratios for this outline are not as close as for the SPO Glyph and the Giza pyramids, still it reflects similar geometry. These relationships may be emphasized by the fact that the antipode point for the pyramids is equidistant from Sajama and the point where the Abbas Giant alignment circle begins to run parallel to the SIO ridge line, thus forming a spherical isosceles triangle with the side lengths ratio of 1:.66:.66, which coincidentally is approximately the same as Easter Island's shape and has the same side length ratios. The graphic below illustrates these relationships.

The Giza pyramids have relatively fewer alignments with other archaeological structures, unlike most of the others we presented so far do. Some of the few alignments are; the diagonal alignments mentioned and a few in the Cardinal directions: immediately north across the Mediterranean in Turkey we find Aspendos and Altinkaya. To the east are the Rajajeel and Junapani circles, which appear to be directional markers. The importance of the Junapani circles is discussed in detail in chapter 12.

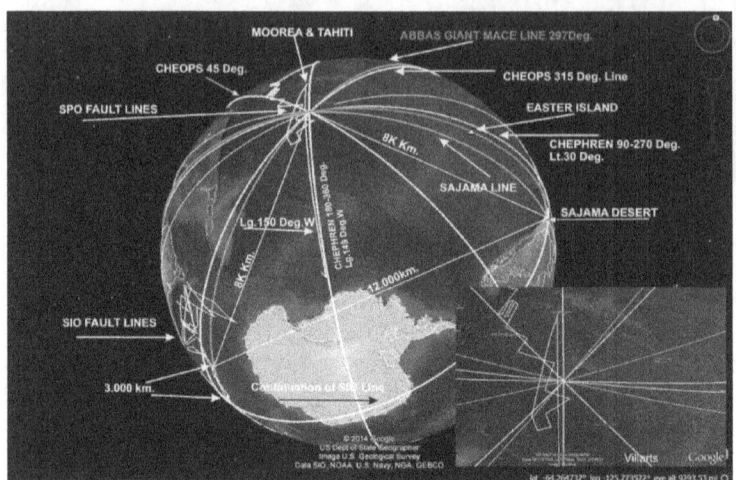

Graphic 4

The geometric alignments discussed here appear to have as their main purpose to set the location of the pyramids on earth; the alignments found for the pyramids answer, for this site, the basic questions that set the objectives for this research: " Why are archaeological structures located where they are found and what determined it?"

From the evidence discussed in this chapter, we come to a conclusion:

The pyramids had as their purpose to rise the geological 'markers' inherent in the ocean plate ridge lines to the surface to be seen by future civilizations, perhaps as a record of the ancient's accomplishments. From these to learn the astronomical facts the earth's plates encode and to serve as gigantic sextants that appear to complete the astronomical connection with the stars, thus placing earth in its universal context. An extraordinary legacy from a technologically advanced ancient civilization.

NOTES

CHAPTER 11

THE CHACO and CHELLY CANYONS
& HOVENWEEP PUEBLOS

 The Chaco canyon Pueblos is a group of about 17 distinct archaeological locations in southern San Juan County, NM (36.08°N 107.95°W). The larger pueblos are Pueblo Bonito and Pueblo del Arroyo. Four miles southeast of Pueblo Bonito is the Fajada Butte. This Butte is naturally aligned with the SPO glyph, as described before.

A distinctive feature of these pueblos is the presence of large circular structures constructed of stone and mortar known as kivas, the largest is found in *Pueblo del Arroyo*, it measures about 72 feet in diameter.

The structure, unlike the others is formed of concentric stone rings at ground level, while most others are tank-like and some are, partially, below grade. This structure consists of three concentric rings connected by radial walls.

The overall design is almost identical to the Sun Dial at Sacsayhuaman, in Cusco Perú, shown on page 77. As with the Sun Dial, its radii point in specific directions: The radius pointing NW has a heading of 325.17°. This is the same heading of one of Nasca's main lines, it may be recalled. A line traced at this angle in the northward direction passes over Chaco Canyon pueblo Yellow House. Further north it cuts through the Hovenweep Pueblo region, connecting the twin towers, the Kiva Pueblo, the Cliff Ruins and Devil's canyon. The Devil's Canyon is, also, mapped by the Ahu Vinapu Ceremonial's 0°-180° line, which on its way north passes over the Chelly Canyon cliff dueler's ruins. In these ruins there are petro glyphs, one of which has a figure that displays four and five fingers as does the Chaco Canyon glyph, shown at the beginning of the chapter. Continuing the 325.17° line north, in Utah finds the Bingham Canyon Copper mine. Around the world the line finds Triangle 2-SIO at its vertex-1, shown at the bottom of the graphic on page 184 and page 238; this is one of the four triangles mentioned earier, it is outlined by the Chaco and Hovenweep Canyon Pueblos. Following the same line on its SE direction at 145.17° the line passes by the Cuatro Palmas gold mine and further down, the El Carmen gold mine, both, now abandoned. Further down, the line cuts through the Gulf of México and in the state of Veracruz passes about eighty miles from the archaeological site of El Tajin, one of the most important Pre-Columbian cities of Mesoamerica. Crossing the Central American isthmus it reaches the Caserones Copper Mine and the Macasin Salt Lake in Argentina. A second line following another radius at an angle of 204.98° finds the Triangle 1-SIO, ending at its vertex-3. The importance of this line's direction is echoed by the pueblo's eastern wall, which runs in the same direction and crosses the Nantack Pueblo ridge. Following the line in the opposite direction at 24.98°, around the world, finds vertex-1 of T2-SIO. This is the antipode point for Pueblo Arroyo and the Chaco canyon region.

Running a line SE from the geometric center of the structure to Nasca at an angle of 142.51º, in Perú, the line crosses over the archaeological cities of Siete Techos (6.84ºS 79.78ºW), Cañoncillo (7.41ºS 79.46ºW), Huaca Arco Iris (8.00ºS 79.07ºW), Cha Chan (8.11ºS 79.07ºW), Sechin (9.48ºS 78.26ºW), Pampa de Las Llamas (9.5ºS 78.22ºW), Pocoto (12.89ºS 76.24ºW), Incahuasi (13.02ºS 76.17ºW), and the Nasca Calendar Wheel (14.64ºS 75.17ºW), before reaching its own line on the Nasca plain which starts at a prominent Nod point (14.698ºS 75.14ºW). Eleven archaeological sites in all are connected by this line.

Another connection of particular interest is with the Inca city of Raqchi, in Perú (14.17S 71.37.W) mentioned earlier. The Raqchi Citadel at one time may have had nearly one hundred Kiva-like structures, locally known as *qullqas* which are believed to have been used for food storage. These are fronted by the monumental Temple of Viracocha; of which today only the sixty foot walls and

National Park Service: Pueblo Bonito Kiva

one column remains. The line connecting these two sites starts at the center of the circular stone structure in the Arroyo Pueblo in alignment with the kiva on the southeast corner of the pueblo. Further south the line connects with two of the Rinconada pueblos, we labeled Chaco 9 &10. The line has a south bering at an angle of 146.56°, in the Yucatán Peninsula crosses over the Mayan city of Palenque and in Perú reaches Raqchi at its center. Continuing south the line reaches the Lake Titicaca region. The line passes on its western shore by the Sillustani Chulpas and the Amaru Meru door; and sixty miles west of Tihuanaco and Puma Punku and continues further south through the Sajama region where round chulpas and kiva like structures are also found.

Pueblo Bonito is a quarter mile east of Pueblo del Arroyo, it has a horse shoe configuration that faces south. It encloses rectangular buildings and twenty five circular Kiva structures. The largest of these circular structures, sits at the front-center and is about 52 feet in diameter and about ten feet deep. This structure is flanked on the east and west by stone walls that run at different angles. The Western wall line runs at an angle of 180.0°, whereas the eastern wall line runs at an angle of 171.17°. The eastern line at 171.17°, reaches the south Indian Ocean, forming the side of the second Triangle, previously mentioned (T2-SIO); connecting its vertices V1 and V2. The 180° line passes 76 miles east of

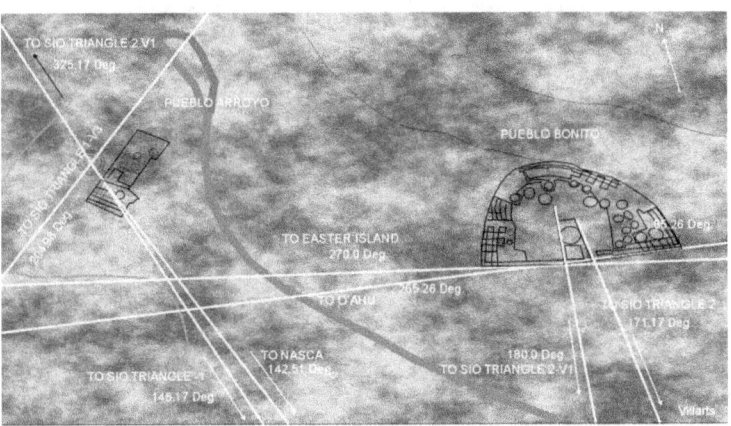

Easter Island and ends at T2-SIO vertex V1; as mentioned before, this point is the antipode of the Chaco Canyon Pueblos.

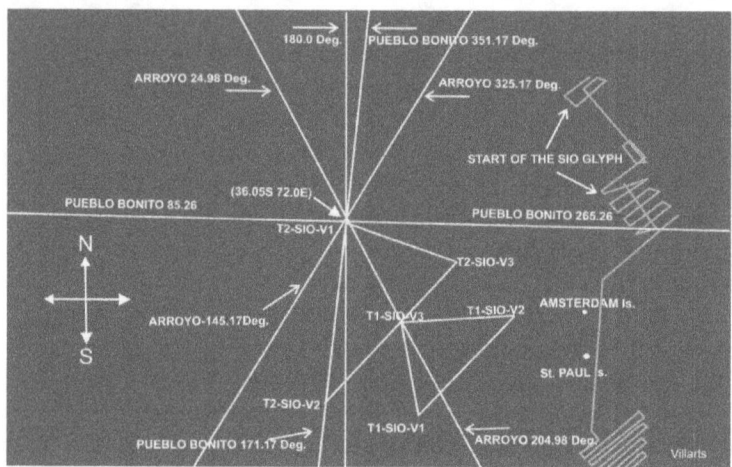

The front of the horseshoe shaped pueblo- see illustration, p.238- is enclosed by two walls; these are not co-linear. The western wall runs east west at 90°-270°. The purpose of this alignment appears to be, the now familiar, setting of the cardinal coordinates, with the 180° western wall from which all other alignments depend. The eastern wall confirms this; it runs east at 85.26°. Running this line west at 265.26°, it reaches the island of O'ahu, passing through Honolulu near Temple Heiau. Continuing on to Australia, it finds the Cascades petro glyphs, and on the west passes by the mines of Laverton and Gwalia, and 167 miles north of the Kalgoorlie mine. The line ends at vertex 1 of T2-SIO, The converging point of the two Pueblo Bonito lines. The 90° east bound line cuts through the Pharr Mounds in Mississippi and Rock Eagle in Georgia.

Ahu Vinapu Ceremonial Center's eastern line has a bearing of 1.27°. A line drawn with this bearing reaches Chaco Canyon's *Peñasco Blanco*. This pueblo's arrangement is semi circular with a radius of 355ft. The line reaches at its center, thus confirming the alignment with the Island. Recall, the back wall of Ahu Vinapu Ceremonial runs at 180°, and ends at Hovenweep Devil's canyon; the 1.27°difference between the sides detarmine the new alignment. These three sites form an spherical triangle with its angles built into their designs and placements; this reveals a planned design. The table on the next page summarizes Chaco Canyon Pueblo's data.

The Chaco Canyon petro glyphs at *Una Vida* (36.03°N 107.91°W) depict glyphs similar to those we are acquainted with: Two spirals and two anthropomorphic paintings; one with a big hand displaying five fingers, while the other figure, who holds one of the spirals on its right hand, has only four fingers on its left hand. The glyph is shown at the beginning of the chapter. The *Chetro Ketl* pueblo's eastern wall has a SE alignment of 155.04°. In this direction, in Acapulco México, it passes near the Palma Sola archaeological site. The glyphs at this site, also, depict anthropomorphic figures with four and five fingers.

CHACO CANYON PUEBLO DATA

	SITE	COUNTRY LOCATION	#KIVAS	Latitude	Longitude
SITE1	PUEBLO BONITO		25	36.061	-107.965
SITE2	PUEBLO ARROYO		10	36.06	-107.961
SITE3	CHETRO KETL		11	36.06	-107.953
SITE4	TALUS		5	36.06	-107.955
SITE5	YELLOW HOUSE		3	36.065	-107.969
SITE6	HUNGO PAVI		1	36.05	-107.929
SITE7	CASA RINCONADA		1	36.054	-107.96
SITE8	RINCONADA 2		4	36.053	-107.959
SITE9	RINCONADA 3		4	36.054	-107.958
SITE10	RINCONADA 4		3	36.054	-107.958
SITE11	TSIN KLESTIN		2	36.036	-107.957
SITE12	PENASCO BLANCO		1	36.081	-108.002
SITE13	KIN BINEOLA		6	36.003	-108.141
SITE14	WIJI		0	36.026	-107.869
SITE15	UNA VIDA		1	36.033	-107.912
SITE16	PUEBLO ALTO		3	36.07	-107.957
SITE17	NEW ALTO		1	36.07	-107.96
SITE18	UNA VIDA GLYPHS			36.034	-107.911
SITE19	PETRO GLYPHS			36.072	-107.912
SITE20	FAJADA BUTTE			36.019	-107.909

81

Also notable is that both figures at Una Vida, have curved antennae strongly resembling the Tirona golden warrior's shown on page 13. These two glyph characteristics will be encountered again.

The picture below is at the entrance of a Navajo historical site in Utah. The glyph depicted is a photographic rendition of a section of the 'Newspaper Rock' glyph wall a few miles north of the Devil's Canyon; note the four toed foot, the curled antennae and the "Calendar wheel"; these glyph pictorial elements are found around the world. The Newspaper glyph location is on the SIO Triangle and A51 Triangle line; shown in the map on page 243.

US Forestry Service

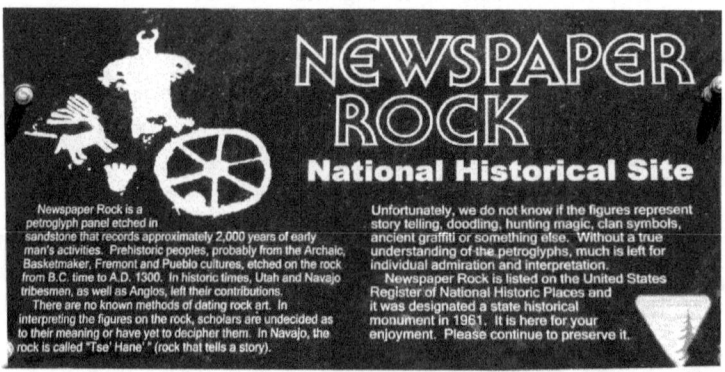

Chetro Ketl's northern wall runs NE-SW at 68.87°, on its easterly direction the line across the Atlantic over Africa finds the Congo-Zambia Copper Belt. This is the fifth connection with this Copper mining site. This line runs nearly the same angle of the earth's ecliptic, as the main line at Nasca does. The line on its westerly direction crosses over the Kin Bineola Pueblo (36.0°N 108.13°W) and in California finds the Blythe Intalgios.

Inside the walls at *Chetro Ketl Pueblo* there are three Kivas that are aligned with the pueblo's north wall. In front of the pueblo,

175 feet away, is a large Kiva. Running a line connecting the centers of the middle Kiva in the pueblo and the large Kiva at 326.62°, in its southeasterly direction it goes over the Fajada Butte and in México it finds the Tontiná citadel. This site, it may be recalled, is the beginning of the Great Arc. In Guatemala the line passes by the Kaminaljuyu archaeological site. Across the Pacific Ocean into Perú it crosses at the center of Mauka Llakta. This Inca Pueblo alone has, nearly, seventy Kiva-like structures compared to 82 in all of the Chaco Canyon Pueblos combined, and the only domed circular structure in South America. This structure has a phallic appearance and aligns with the phallic megalith in China with a line that runs at an angle of 355.27° and at 354.37° with El Infiernito in Colombia. The line connection between the megaliths in Colombia and China runs at 360°.

THE HOVENWEEP PUEBLOS
This Pueblo site has been extensively studied, in particular for its astronomical solar calendar window arrangements. The windows allow sun rays to illuminate different wall sections at different times of the year. One of these illuminated walls displays the, now familiar, spiral. The Pueblos are found in the canyons around the 'Four Corners' where the States of Utah, Colorado Arizona and New Mexico meet. These pueblos are geometrically connected with several archaeological sites around the world. Some of these connections have already been explained, specifically, those of the Chaco canyon Pueblos. Following we focus on the Hovenweep structures and their alignments. We start by showing the geography of the region on the map in the next page: the Four Corners region. Hovenweep is just north of the four corners, which are marked on the map with dotted lines.

The Chaco Arroyo 325.17°NW line passes two miles east of the Hovenweep Twin Towers and the Castle. The Twin Towers (37.37N 109.07W) are aligned at an angle of 140.2° following this direction south to Nasca, the line finds the Nasca line for the towers at 14.68°S 75.10°W.

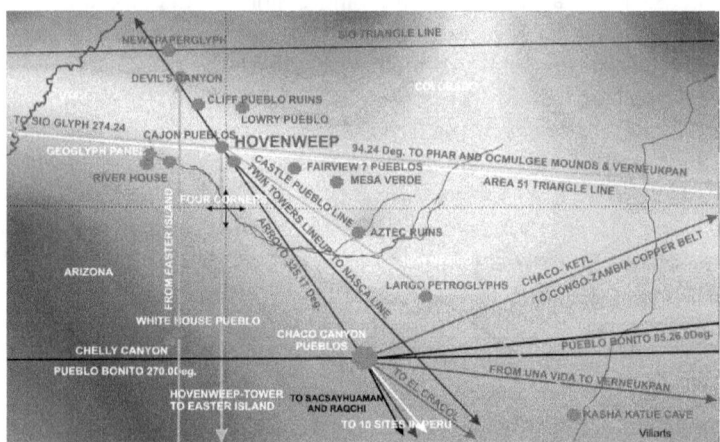

The larger tower's inner wall runs southward at an angle of 132°. In this direction, in the Yucatan Peninsula, Mexico it finds the pyramid Citadel of Ezná (19.6N 90.23W), the Calkamul Pyramid (18.86N 89.51W), The Lamanai 'Mask' Temple in Belize (17.67N 88.65W), the Salutara petro glyphs in Nicaragua and the San Agustin archaeological Park in Colombia (1.92N 76.78W). The half round Tower (37.38N 109.08W), west of the Square Tower, has a straight east side wall which runs north-south at 180.32°.

Following this line south to Easter Island it finds Ahu Vinapu Ceremonial. The dividing wall within the structure is nearly perpendicular to the back wall with a bearing of 274.24° in this direction the line finds the Cajon Group pueblos. Continuing west, it crosses atop the Area 51 Fork glyph and in the south Indian Ocean crosses the SIO Glyph and the T2-SIO-V3.

In Africa it passes thirty two miles north of Verneukpan and forty miles north of the Black Mountain Copper mine. On its easterly direction the line finds the Pharr Mounds in Mississippi and the Ocmulgee Mound in Macon, Georgia. This mound which has a flat pyramidal shape is aligned with the ball court at Chichen Itza, which has a bearing angle of 17.95°.

The Square Tower is aligned at 107°. Following this line southeast it crosses over the Reservoir at the Anazasi Fairview Pueblos. The eastern wall is aligned with the White House Pueblo at Chelly Canyon and the Petrified Forest. The Castle Tower south wall has a bearing of 140.2°. In this direction, in New Mexico, finds the Aztec ruins, the Largo petro glyphs and Tapacito ruins. Continuing further south the Bandelier Cliff Dwellings and the KashaKatue cave are encountered.

In the Yucatan peninsula, the line passes near the Nohoch Mul Temple at Cobá (20.49°N 87.72°W) and the Muil ruins (20.16°N 87.56°W) and in Colombia goes over the salt mines at Zipaquirá, thirty miles north of the El Abra archaeological cave.

CHAPTER 12

MENHIR STRUCTURES

CIRCLES

In Chapter one, we spoke about the importance of comparatively smaller structures in the archaeological landscape, such as dolmens, mounds and circles or wheels. We found in this study some of the menhir circles serve two functions: they are, commonly recognized as astronomical event tracking structures and we found they are geographical markers, as well. Their astronomical function has been studied for many years; those studies have shown that their alignments were useful to predict the phases of the moon, the position of the sun, as well as, the solstices. In this study we found that their geographic location, usually serves as central points for the alignment of two or more locations, or to point out the direction to other archaeological sites. These two properties combined; appear to have served the same purpose a Sextant would have for a mariner who uses it to avail him of the stars for guidance. In chapter 6 we described how the geographical positioning and alignment of some of these stone structures marked a specific arc degree of rotation of the earth in front of the sun, indicating the solar time of day.

STONE ARRANGEMENTS

Some of the stone circles are basically circular; however a large number of stone arrangements are not perfectly circular although they may enclose an area. They can be oval, oblong, rectangular or linear -a narrow double row with a few end pieces or a dolmen or end arrangement. We start this discussion of 'circles' with the linear type.

The Tomb of the Giant in Lunamatrona in the Island of Sardinia is an example. It is particularly interesting due to is close alignment with the Circle of illumination. Its azimuth is 35°, whereas the circle of illumination at that latitude has an azimuth of 31.09°.

At the summer solstice on June 21, past the culmination at 1242 hrs the sun shines at an azimuth of about 145° with an altitude of 71.5° thus it barely shines through the 'forum' or circular hole

on the head stone. However at the winter solstice on December 21 at 1425hrs the sun has an azimuth of about 145° and an altitude of 18.41°, thus it illuminates through the forum the length of the Tomb. The Tomb of Madau $(40.120704^\circ N\ 9.331885^\circ E)$ is another interesting arrangement. It is the last dolmen in the alignment with Stonehenge's Circle of Illumination at the summer solstice, which was described in chapter 6, p.145. It may be recalled that alignment consists of the dolmens of Grotta, Chiaramonti, San Nicola, Sa Conca e s'abba, Orotelli, Mazzozzo, Mamoiada, and Tomb of the Giant in Madau. The Tomb's structure is aligned lengthwise at an azimuth of 148.64°. At 1242hrs on June 21 the sun shines at the entrance at an azimuth of -31.36° and an altitude of 71.1°

THE MEDICINE WHEEL

During the course of this research we have come across the Medicine Wheel quite often. The repeated encounters inform us its location on earth was particularly important. For the peoples of the region it had various powers including, of course, medicinal ones. Its importance today is still celebrated not only as part of the history of the native inhabitants of the region, but as a site of some mystical communion; a Stonehenge-like fascination although its structure is as simple as it could be made; but, why?

It turned out its importance as a geographic marker may be as important as Stonehenge and equal to other rings that will be discussed below.

Early on we encountered the alignments between the Medicine wheel with the Nasca Calendar circle and the Caracol observatory with the Sacsayhuaman Sun Dial line, both closely follow the angle of declination of the earth's axis. The Medicine Wheel-Nasca Calendar circle alignment is significant in that it is only 0.65° off the angle of the Circle of Illumination of 145.87° at the summer solstice at the Medicine Wheel's latitude of 44.82° The Medicine Wheel shares the Circle of illumination with Nasca and the Angkor Wat temple in Cambodia due to that small difference in angle. A circle at that azimuth (145.19°) connects the three sites precisely. Similarly, the winter solstice Circle of illumination is shared with the Ring of Brodgar and Giza. The corresponding circle with a small azimuth difference passes thirty four miles off shore the Ring of Brodgar and reaches Giza over the Pyramid of Chephren.

The great circle which connects the Medicine Wheel and Stonehenge aligns over the Hill of Tara Stone in Ireland and passes thirty miles off shore of Tahiti where the Tevaifaara stone is found; both stones are phallic megaliths. The Medicine Wheel is also in alignment with the phallic megalith giant at El Infiernito in Colombia. Southeast of Colombia, in the Estate of Rodônia in Brazil the same alignment finds the petro glyphs of Rio Madeira.

Going northwest, off the coast of Ishigaki-shima the circle crosses over the Yonaguni Pyramid.

The Medicine Wheel's other alignments with other rings will be described as those ring's alignments are discussed.

THE CALLANISH RING

The Callanish ring is a stone arrangement consisting of a circular ring bisected north-south by a double row of megaliths to the north and a single row to the south. The eastern row has an azimuth of about 192.12°. The western row has an azimuth of about 191.92°. A circle drawn at an azimuth of 192.12° reaches one of the stone circle arrangements of Sine in Senegal. A second circle following the western alignment at 191.92° reaches the Wassu circles arrangement about forty four miles east of Sine. At Wassu the circle aligns with three of the rings of which the middle one has two standing megaliths similar to those at Callanish.
The east-west megalith row runs at an angle of 264.35°, a great circle at this azimuth reaches Hanga Roa beach in Easter Island. Two of the megaliths on the circle align with the center one at an angle of 244.11°. A circle drawn at this angle reaches the Nasca plain where it finds a polygon with this alignment (14.781648°S 75.177559°W). Another circle aligned at 81.94° points to the stone circles at Junapani, India and one at 156.13° reaches the spirals of Verneukpan in South Africa.

SALBYKSKY MOUND RING
This burial site was excavated between 1954 and 1956 by MSU professor S. Kiseljov (unknown Siberia.com). It has an oblong shape of about 70X70 mts. built with stone slabs weighing about 50 tons each, it is located in Siberia at 53.894°N 90.773°W. Its northeast southwest alignment diagonal across the quadrangle at an angle of 217.12 closely aligns with Easter Island. A great circle in that direction cuts across the Island connecting Ahus Ko Te Riku, Tahai and Vinapu. Earlier on pages 123-4 we discussed this mound's alignment with the Stupas of Polonnaruwa.

Polonnaruwa is aligned at 7.88°, a great circle at this angle crosses over the mound and continuing over the North Pole, further south aligns with the Medicine Wheel.

THE CROMELEQUE

In Xerez, Portugal there is another oblong menhir arrangement with a giant menhir at its center. The western menhir alignment has an azimuth of 8° north, the eastern alignment's azimuth is 18.49°. South of the quadrangle, at 188°, in Lavajo it aligns with two menhirs of that name. West of the cromeleque at 281.13° a line nearly perpendicular to its north alignment finds the Cromeleque of Almendres. This arrangement has a double circle; the 281.13° line is tangential to both circles' menhirs on the north side. At Xerez, the northwest and southeast corner menhirs align with the central giant menhir at an angle of 329.47° which is the angle of the summer solstice Circle of Illumination at this latitude. This circle of Illumination is shared on its southeast direction by the Dolmen de la Hueca, the petro glyphs of Tajo and the petro glyphs in the Cueva de las Palomas in Spain, across the Mediterranean into northern Africa in Algeria finds the petro glyphs of Taghit and the ruins of Ksar Amguid. In Nigeria near Mout St. Yves finds the Shere Hills Cairn and the Holes in granite. Further south in Cameroon, the Rock Ndonkol also aligns. Continuing north after crossing Antarctica in the Island of Tonga the circle finds the Ha'amonga Trillithon, in Samoa passes forty miles west of the recently discovered sunken 4k BC year old city of Mulifanua in Upolu.

The northeast southwest corners of the cromeleque align through the giant menhir at the center at an angle of 238.79°. Drawing a circle at this azimuth it reaches the Temple of Kalasasaya in Tihuanaco, and aligns over the Ponce stela in the center of the quadrangle, continuing southwest it crosses over the city of Puma Punku.

SENEGAL STONE CIRCLES

The Senegal stone circles site is located northwest of the town of Sine (13.695°N 15.535°W). It is one of two sites about 44 miles from each other. The other site is in the town of Wassu. Both sites contain numerous circles, the total number is unknown some of the sites remain still unexplored. They differ in design; the Sine site -shown below- shows how the circles are scattered in no particular design, although the general direction points northwest, the other site (the main known circles) are geometrically aligned.

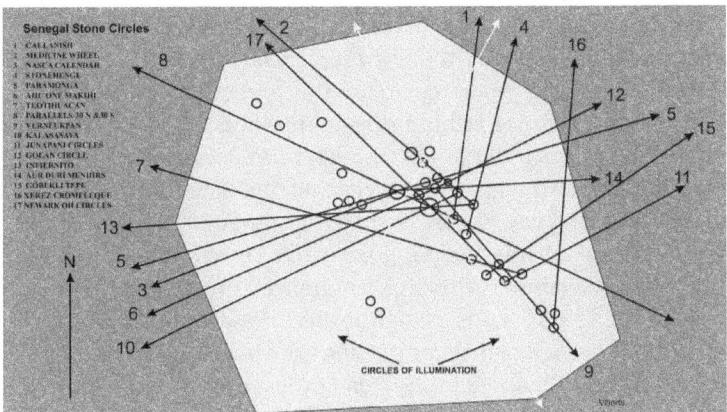

The Sine site is fairly unique in design, it consists of many small circles in apparent disarray; the largest circle is nine meters in diameter. There are other sites with similar layout; one is in Junapani in India, the other is in the Sajama desert. Junapani is similar in design and coincidentally has the same number of circles -or close to it. As we will see that design also follows a general direction. Sajama is vastly larger and its circles are not menhir arrangements but Kiva like structures easily confused with animal corrals. The circular structures are scattered over an area of over twenty thousand square kilometers vastly larger in size over the others. These structures also appear to be in total disarray; however they also tend to align in one direction. The Sine Senegal lines roughly follow the summer solstice circle of illumination, while the Junapani circles are aligned to the winter solstice circle of illumination. The Sajama circles' alignments follow the winter solstice circle of illumination in the southern hemisphere. The disarray is only in appearance for the three sites; as with the Nasca lines, the alignments of various circles within each site align with other locations with circles or archaeological structures.

The first alignment of rings we find at Sine is with Callanish. The alignment is reciprocal, see graphic, alignment number (1); as described above, the Callanish eastern megalith row points south to the Senegal circles at 192.12° there it aligns with two circles.
The second alignment (2) is that of four circles arranged at an azimuth of 314.38°, a great circle in this direction, in North America reaches the Medicine Wheel. This alignment is also reciprocal; one of the Wheel's radii through cairn B has an angle of 55.87°, at this azimuth a great circle reaches the Sine four circles on the same line. Two circles forming alignment 3 have an azimuth of 246.59° at this angle a great circle reaches the Nasca Calendar circle. Alignment (4) with Stonehenge is formed by two circles aligned at 13.63°, a great circle at this azimuth cuts through the center of the Henge and aligns passing between megaliths: 36-37, 54-55b, 106a-160b and 28-29; making this another reciprocal alignment. Alignment (5) is made up of three circles aligned northeast-southwest at 71.7°-251.7°. A great Circle at this azimuth finds the Fort of Paramonga in Perú. This fort is believed to be of pre-Inca, Chimú origin. Following this direction to India the Subrahmanyeswara Temple is found. Following the circle on its northeast direction across Africa near the Nile River, it aligns with the Nabta Playa Aswan Circle and passes fourteen miles north of the Abu Simbel Temples. Ahu One Makihi on a northeastern direction of 79.39° points to the circles at Sine. At the site a great circle in this direction aligns (6) with two circles in it. Following the circle northeast, in the Nile valley finds Hermopolis and Antinopolis. Further east across the Arabian Peninsula finds Moenjo Daro in the Indus Valley. In Maynmar crosses close to the Mount Popa Temple and in Cambodia the

Stupas of Wat Tham cave. Over the Pacific Ocean in Perú finds the ancient burial sites of Colca and Tianjani. The alignment of two circles (7) at the site, with a bearing of 287.87°, points in the direction of the Caribbean Sea and México. In the western side of Cuba, a great circle in this direction finds the sunken pyramids at Guanahacabibes, crossing the Caribbean Ocean onto the Yucatan Peninsula it crosess by the pyramids of Dizibilchaltún. Across the Gulf of Mexico the circle reaches Teotihuacan and crosses over the Pyramid of the Moon.

Alignment (8) is unusual, it appears to have no other purpose than to connect the 30th north and south parallels; however it connects them at the 80.48° and 100.48° longitudes. This is significant in this study because the Aur Duri menhirs are located at 0.038°N 100.48°E. It may be recalled 100.48°E is the Time Zero longitude for the Ancient Global Clock discussed in chapter 6. The Aur Duri menhirs are also aligned with the Sine Circles (14). Another fact which adds to the importance of the point of tangency is the fact that the earth's axis is drifting along these longitudes; down on 80°W, up on 100°E.

This is one of three alignments that are formed with the largest circle, which emphasizes their importance. Also it is the third great circle which connects these parallels; we reviewed the two others earlier: Ahu Akahanga which connects the parallels passing over the Pyramid of Cheops and the Verneukpan spirals arrangement which starts at the 30°S parallel and its inverted 'V' ends at the 30°N parallel. Alignment (9) formed by three circles reaches Verneukpan at the vertex of the arrangement on the 30th parallel. Alignment (10) is the second alignment through the largest circle. A great circle drawn through the center at an azimuth of 241.7°, crosses over the Kalasasaya Temple in Tihuanaco and over the Ponce Megalith at the center of the plaza. Alignment (11) connects the Sine Circles to the Junapani Circles in India. These two sites are similar; both contain multiple stone circles of different sizes. At the Junapani site a great circle from Sine with an azimuth of 68.36° aligns with two of its circles; it is a reciprocal alignment. Alignment (12) connects the Sine circles with the

Golan Heights Circle. Alignment (13) is the third alignment through the largest circle. At an angle of 268.02° a circle in that direction finds El Infiernito in northern Colombia. At the Infiernito Archaeological Park the circle lines up with four megaliths; two in the main circle and two in the smaller circle (5.647 °N 73.559°W). This is another reciprocal alignment. The alignment with Göbekli Tepe (15) is also a reciprocal alignment. A great circle with an azimuth of 53.75° formed by two circles at Sine reaches Gobekli Tepe and aligns with menhir 16 in enclosure B and menhir 37 in enclosure C. See graphic on page 112. Alignment (16) is also reciprocal; two circles at an angle of 18.49° align with the eastern side of the quadrangle of the Xerez Cromeleque. Alignment (17) points to the mound circle of Newark Ohio in the US located at (40.41°N 82.4311°W).

THE JUNAPANI CIRCLES

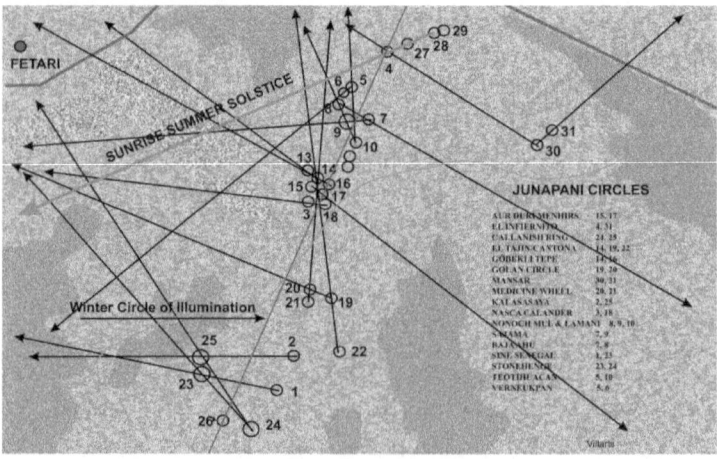

These circles are located at 21.199°N 79.00°E, and as mentioned earlier the circles are scatterd but they follow a general direction; they follow the winter solstice circle of illumination. It is shown in the graphic above. It is the line shown going through circles 4, 16, 17 and 26. In the graph the direction of the sun at sunrise at the summer solstice is also shown (4,27,28,29).

The Junapani circles align reciprocally with: Callanish, the Medicine Wheel, the Sine circles, the El Infiernito phallic menhirs circle, Göbekli Tepe and Stonehenge; that is they point to each other with an alignment in the design of the structures within each site. For the other alignments shown in the graphic, although the direction of the alignment points to the location of the structure or citadel, a great circle drawn at the angle of the alignment reaches the location but does not align with any feature of the structure. Some of these alignments could be thought of as serendipituous, but, in some instances the alignment is confirmed by the multiplicity of structures at other archaeological sites that are found on the same path. This is the case of the alignment of circles (22, 19, 14) headed north over the North Pole and crossing over North America into México. Staring near the coast of the Golf of México the alignment finds the Citadel of El Tajin (20.447°N 97.376°W) crosses over two of its pyramids in the direction of its main alignment and continues southwest passing near the Pyramid of Yohualichán (20.062°N 97.503°W), then past the largest Mesoamerican Citadel of Cantona (19.552°N 97.488°W) and further south near the pyramid of Teteles (18.627°N 97.717°W). Similarly the alignment of circles (8, 9, 10) points to the Yucatán peninsula in México, a great circle in that direction reaches the archaeological site at Cobá passing by the pyramid of Nonoch Mul (20.493°N 87.721°W), then, continues southwest crossing over the pyramids at Chaccoben (19.001°N 88.230°W). Across the border over into Belize, the alignment finds the Cerros pyramid (18.359°N 18.359°), the Lamanai Temple (17.767°N 88.652°W), the Actun Tunichil Muknal sacrificial site (17.117°N 88.890°W) and further south the Belize Caracol (16.764°N 89.117°W). This last alignment of archaeological sites on the eastern side of the Yucatán Peninsula was mapped earlier from

Easter Island and Göbekli Tepe. It is worth noting that this alignment is one of two alignments which go through Junapani's largest circle (9). The other is the alignment of circles 7 and 9 which when followed around the globe finds the Sajama desert. Recalling; the Sajama region covers nearly seventeen thousand square miles. And as mentioned earlier it is believed to have been the site where an ancient civilization flourished, perhaps as early as ten thousand years ago. Of the (se) civilization(s) remain a myriad of intriguing lines similar to Nasca's of which we have cataloged the points where more than two intersect and we refered to them as nodes, some of these nodes have at the intersections what appear to be building structures or pueblos (the locations of 107 of these are given in the apendix) in addition we described the presence of numerous structures similar to the North American Pueblo Kivas which are also found in Peru in cities like Mauka Ilakta. In Sajama we have identified and cataloged over one thousand such structures, the locations of few of these are also given in the apendix. A large number of the 'kivas' in Sajama are stringed together in groups which are aligned between 335.05° to 350° degrees which closely matches the angle of the Circle of Illumination of 335.09° at their latitude. The earth's obliquity angle has been oscillating ±1.3° from its average of 23.3° according to the Washington State University, and has tilted as much as 24.2° which according to their calculations it reached this maximum 9500 years ago. Perhaps, not coincidentally it coincides with the kivas alignments at about 335°; the civilization at Tihuanaco, just north of Sajama, which was believed to have existed about 15,000 years ago by Arthur Posnasky and whose theory has been largely rejected by most archaeologists. We cataloged about twenty seven such structure strings but believe this represents a small percentage of the total. Also, we have identified four regions where 'lost cities' may have existed. These four 'Lost Cities' are intersected by this alignment. The first region covers seven thousand square miles, it is found at (18.291°S 67.252°W). The second region covers ten square miles (18.307°S 67.482°W). The third region covers one half square mile (18.391°S

67.830°W). The fourth region covers seven thousand square miles (18.520° 68.560°). The alignment also crosses by nodes: 32 & 29.

The foregoing discussion of stone circles and enclosures is not exhaustive by any measure; there are literally hundreds of known stone circles and enclosures. As we have seen in many instances they serve as directional markers; this we found to be more prevalent, for the simpler 'circular' structures which seem to be part of a group built for this purpose. Of this group there are five sites which seem to have as a secondary purpose to serve as location markers disregarding their other arrangements; a sub set which includes the Medicine wheel, the Junapani circles, the Nasca Calendar Wheel, the Nabta Playa circle and the Hanga Te'e circle.

All these circles are connected in a structured arrangement. The lines connecting them converge at the SIO Triangles. The connecting line between the centers of the Medicine Wheel and the Junapani Circle, from the north, passes over the Jageshwar and Kahuraju Temples in India. Continuing the line to the south, it passes over the Kolar Gold fields in southern India. Continuing around the globe back to the Medicine Wheel, the great circle passes through V3 of T1 SIO.

Likewise, the line connecting the Junapani circle with the Nasca Calendar wheel passes over the Dholera Swaminarayan and Kayavarohan Shiv Mandir Temples in western India and in Egypt crosses by the Abu Simbel Temples, and the Nabta Playa Aswan menhir circle.

The line that connects the Medicine Wheel with the Hanga Te'e circle, in Easter Island, on its way south passes over Kin Bineola, one of the Chaco Canyon Pueblos, and in the South Indian Ocean reaches the V1 of the T2 SIO. The Kin Bineola pueblo is located eleven miles outside the group of pueblos in the canyon. This location appears to have been selected to be specifically located in the Medicine Wheel-Hanga Te'e circle alignment path. The line connecting these two circles has a bearing of 181.35°. The

Medicine Wheel's location is 0.16°, due north the Pueblo Bonito, and is connected to it by its 180.0° western wall line. Again, the 1.35° differences in angle appear to be designed. The Kin Bineola pueblo contains five kivas. The network of circles just discussed contains five circles, as well: The Medicine Wheel, The Junapani main Circle, The Nabta Playa Circle, The Calendar wheel and the Hanga Te'e circle.

MOUNDS- Archaeological mounds are another type of structure whose alignments seem to depend mostly on geographical positioning, rather than geometrical pointing. The importance that kind of alignment has is exemplified by the alignment of mounds between Scotland and Mexico. Included in this alignment are: the serpent mounds at Loch Nell Scotland, the Ohio Serpent, the, the Newark earthworks (mapped by the Sine Senegal circles 17) two miles away from the Newark, OH circle. This mound consists of two rings: a circle and an octagon with their centers aligned at 51.04°. A great circle at this angle crosses over the Athenean Acropolis and in Saudi Arabia over Moses Altar and the Narjan petroglyphs. The Noch Nell alignment continues past the Pharr mounds in Mississippi, the King's crossing mound in Louisiana and across the gulf into Mexico where it finds the Pyramids of Corregidora and Ihuatzio archaeological sites. Both pyramids are mounds; the one in Corregidora is known as "Cerrito" in Spanish Small Hill, both are shown on the next page. These last two archaeological sites are mapped in the Judaculla Rock.

Corregidora El Pueblito

Ihuatzio

TRELLEBORG

The Trelleborg is a henge type ring believed to have been used by the Vikings as a fort. As pointed out earlier, this ring is found by one of the alignments of the Ahu Ko Te Riku. The entrances to the ring are aligned at 10.8° and 100.8°. At the 10.8° angle, to the north the Barrow at Farubo and the Vejrhøj mound are found. Outside the ring, arranged in radial fashion are the vestiges of 13 canoes shaped stone arrangements of the same kind as found inside the ring which are arranged into quadrants. These are believed were foundations for buildings in the fort. A few of these foundations have alignments to remote locations. The first one at an angle of 331.61° three miles away finds the Breddysse Dolmen, a second alignment at 326.15° aligns with the Fyrkat fort and Aggesborg. The third alignment reaches the Medicine Wheel in Wyoming, US. The fourth alignment is with the Caracol Spiral of Texcoco, México and the round pyramid of Cuicuilco near Tlalpan in México City. A great circle following this alignment at 297.44° passes four miles from Teotihuacan - the Sun and moon Pyramids and two miles from Tenochtitlán (México City' Zócalo) the capital of the Aztec civilization.

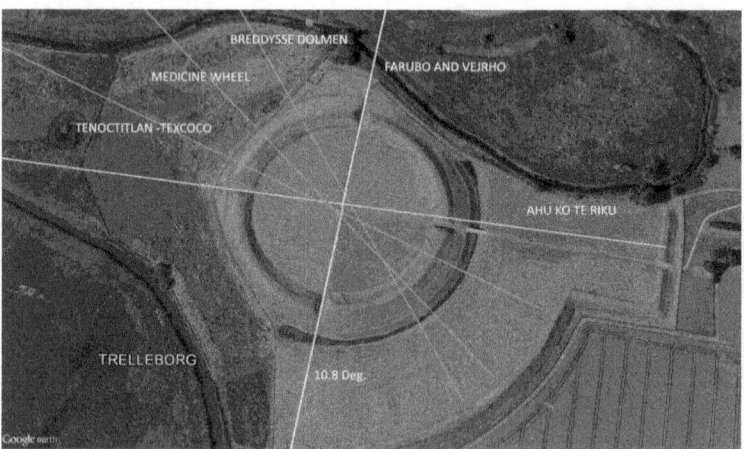

CHAPTER 13

EPILOG

This work began by proofing that ancient structures in the Middle East, are aligned in specific geographical ways; as was claimed by writer Zacharia Sitchin. We also verified the work of Sir Alfred Watkins the 'father' of line alignments in the UK (Lay Lines); then extended the scope of the study globally to find out if these were unique occurrences.

This research uncovered important facts about the geographical location and the geometric alignment of well over one hundred ancient archaeological structures, and their location's relationship with natural sites. A significant number of redundant data points were collected for some of the alignments that together reveal the, up to now hidden, but, openly measurable global framework made up of structures which connect with our earth's geology and with the points where our civilizations flourished.

We found there is measurable global coherence in the location of archaeological sites. The structures are aligned in great circles. Most of the circle's azimuths for each location were found to be mapped in the geometrical features of places like Nasca and Easter Island. The location of many archaeological sites is triangulated from other sites which point to them; their circles intersect at the given archaeological site. Triangulation of coordinates for a location virtually eliminates the possibility, for that location to be found in a particular place by chance. For many of those sites which were not triangulated and did not have two lines crossing at the same coordinates, their positioning was justified by other factors: most prominent their locations are aligned with the circle of illumination; they receive sun light at sunrise simultaneously, regardless of their latitude. Also, structure type, design characteristics or some other pertinent features were used to confirm the alignment. Still, some sites may be on a line's path due to coincidence alone as can be expected. This fact in no way negates the overwhelming evidence, when it is evaluated as a whole, especially in instances where the purpose of the alignment becomes obvious.

No attempt was made to attach to the data any accepted chronology of geological or human development. Our focus was on the immediately measurable physical evidence; also, without attempting to explain how the record was created, this resulted in a pragmatic, factual and reproducible cartographic account of the geometrical alignments found. Likewise only a few speculative conclusions are made; following are suggestions as to how these discoveries may be useful. The discoveries made during this research will help practitioners of the various sciences that focus on the history of human development to define or solidify the chronology of events, by firmly connecting seemingly unrelated regions or archaeological sites and their societies, i.e. Easter Island and Giza. Associating sites and the history of the peoples for each era, using the given alignments as an overlay, may help correlate the history of the various global regions.

This may encourage some researchers to shed off the traditional regional view for studies on human development. For example, we found, a geographical correlation of sites where the *practice* of elongating the human skulls existed. Some of the sites are in western México: Onavas, Sonora, Marismas and Nayarit, in Colombia; Chicamocha, in Perú; Paracas. They are, also, found in other countries such as in the US, Ukraine and Egypt. Egyptian art known as the Amarna depicts the royal family with elongated skulls. The geographical diversity of these locations where the skulls are found is a strong indication of the universality of this phenomenon. This hints at the possibility of an ancient human race having existed with this skeletal peculiarity. This would make the paintings and sculptures of that time period realistic depictions of those peoples, not art and skull elongation not a ritualistic practice as is commonly held today. This we think has a strong possibility of being the case.

If only for a moment, we contemplate the idea of an alien race visiting earth, it would explain, not only the skulls but many of the customs and attires our societies have shared and still share. There are various examples of customs that developed amongst the peoples of a region as they mimicked their leaders 'style' in usage or speech patterns.

One example is the 'th' sound for the letter 'Z' in Spain specifically in Castile. The legend says the king used to have a lisp, therefore the population at large lisped on purpose to be 'in'. With this in mind we can easily imagine our ancestors observing the shape of our visitors' heads, believed they were deities and tried to emulate this feature by head boarding their children's skulls into that shape, perhaps with fatal results. It appears that many of the elongated skulls that have been found are those of children. Now with respect to attire, in this research we pointed out the many glyphs that display ant-like antennae and helmet-like heads with round eyes. If we look at various cultures' headdress styles, many of them could be local interpretations of helmets with antennae. The most blatant examples are the emperors' crown or the chieftain's feathered headdress, and everything in between that would simulate, not only the physical objects but their meaning: power. Earrings could be interpreted as depictions of wireless headphones. Good examples of this are the Olmec megalith heads; they have both helmets and earrings, the same as some of the moais do. The round eyes found in many glyphs could be goggles...

Another skeletal peculiarity, that was pointed out in the study, are the depictions that show, possibly, the same human-like individuals with four fingers on one hand or foot and five on the other. The most obvious of these depictions are shown in the 'Hand' and Monkey glyphs on the Nasca plain. Others were, also found at: Easter Island, The Newspaper petro glyphs in Utah, the Chaco and Chelly Canyons, in Arizona and New Mexico, US and the Palma Sola glyphs in Acapulco, México. There are no reports of mummies with four fingers; perhaps due to their fragile nature.

The purpose and utility of the Nasca lines and Easter Island's placement of Ahus and Moais was established. We followed these 'maps' and found that the directions they point to, were systematic and geometrically defined. These results lead to the conclusion that the sites served to record a survey of the earth and the location of its inhabitants. We also found other sites with a similar function; all these belong to a category of, 'stars' similar to the ones found on the Nasca plain, once their survey maps are completed, i.e. extending their lines. The best examples are El Infiernito and Avebury; compare their graphics on pages, 109 & 207. Others are; America's and UK's Stonehenge, the Area 51 glyphs, Verneukpan, and the Abbas Giant. Also, the stone circles like the Medicine Wheel, the circles of Sine and Junapani. There are other sites that remain to be more thoroughly explored such as the Sajama desert lines, for which we identified over one hundred Node points for the 'star' lines, also thousands of Kiva like structures and the strings of these which align with the declination angle of the earth's axis at the summer solstice in the southern hemisphere; and also the vast hidden settlements of probably ancient civilizations covering at least 17,000 square miles. Most sites record in their structures astronomical facts of one kind or another. The designers of these places, may have developed, or assisted in the development and placement of these immutable landmarks: the pyramids, megaliths and citadels. The location for these monuments were not only chosen, but also, was the positioning of each monument within each site. Their latitudees and longitudes were precisely set as reported by various researchers. We verified several of them in this study and also found alignments that were built into their design geometries, which in many cases were confirmed to be reciprocal with other monuments' alignments which in turn point back to them.

The flight capability of the ancients was confirmed via the record they left from their exploration of the oceans; they were able to view the earth's ridge lines (glyphs) which they copied and engraved on the mountains and valleys around the Nasca plain. The ocean ridge lines were used as reference landmarks; some alignments appear to coincide with the direction of some of the ridge lines. The great circle which connects the Abbas Giant with Giza follows the diagonal line of the base of the pyramid of Cheops. In the South Indian Ocea finds the ridge line and follows it for nearly 1,400 miles. The Great arc which connects archaeological sites from Abu Dhabi to Tontiná in México, the continuation of the circle aligns with a different ridge line part of the same SIO ridge line for 1,500 miles.

In chapter ten we explored the likely rationale for the Giza pyramid's location, bearing, alignment with each other and their relative sizes. We demonstrated how these parameters closely reproduce the measurements and layout of a second ridge line in the South Pacific Ocean south of the islands of Tahiti and Moorea. The main line of this ridge line runs down to Antarctica pointing directly south for six thousand miles. The pyramids of Giza align perfectly north parallel to this meridian line 150°W.

The pyramid's alignment north south with this six thousand mile ridge line lead us to conclude the pyramids' bearing north was derived from this line. Consecuently, the "North" concept was derived from this ridge line as well, not the mobile Polar Star.

The comprehensive analysis of the South Pacific Ocean ridge lines showed its pattern was reproduced and coded into the design of the great pyramides' site as well as the pyramids dimensions. That part of the study strengthens our knowledge of the degree of technical advancement the architects of these monuments possessed.

In general, it was determined and demonstrated that geographical alignments, of all kinds, between sites, sites and glyphs, glyphs with glyphs are relatively common place and that some of them are tied to geological formations or topographical landmarks, whether they are found on the continents or in the oceans. It was shown these alignments, regardless of type, are not limited by distance, or interrupted by the earth's topography. In the microcosm of Nasca and Sajama, their straight lines run for miles, regardless of the immediate topography, whether found in mountain, river, desert or dale. Does this mimic what was observed with ridge lines on the ocean floor, which span several hundred times the distance the others cover on land?

Themes were found in the study which establishes groups of sites that were determined share a common purpose based on their design. These sites, it appears in some instances, not to have any other purpose but to serve as compasses. This characteristic is unlike any of all the other archaeological sites surveyed. These sites have glyphs or non-inhabited structures within them that serve to document the location of other sites. In order of importance, measured in terms of the number of sites that were determined they lead to, as of this writing, they are: Nasca, El Infiernito, the Ahus of Easter Island, the Medicine Wheel, Stonehenge, the Sine Circles, the Junapani circles, the area 51 Glyphs, Verneukpan, the Candelabra, the Abbas Giant and the Atacama Giant.

Another set of sites that have similar geodetic properties is a group of inhabited structures with alignments not to the cardinal points, but point in specific directions with an established purpose. Some of these are: Teotihuacan, the Chaco Canyon and Hovenweep Pueblos, the Giza Pyramids complex, the Athens Acropolis, the Apollo's Temple at Rhodes, the Jupiter's Temple at Baalbek, and the Temple on the Mount, Puma Punku and America's Stonehenge.

To these groups we add another made up of, single, isolated stone circles or menhirs and other geometric glyphs, by themselves or as petro glyphs, which also serve as guideposts. Of particular usefulness are the circular, quadrangular or triangular markers arranged in a manner that serve as directional pointers to important archaeological sites. Some of these geometric features were, found as parts of structures at a site, or within the site; such as in pueblos or citadels.

It was discovered that the main Nasca lines divide the world in unequal quadrants. The NE quadrant was found to contain the overwhelming majority of the archaeological record. It was also determined that most sites' locations are grouped in steps with the parallels. It was noted that vast areas of the globe, above and below the 50th parallels have minimal archaeological record. However, the few markers that exist are in alignmenmt with other sites from various directions; the Inukshuks are notable.

It became readily apparent that no record stands alone, unconnected to anything else. We showed that some design characteristics replicate globally, sometimes in a small detail, such as: the infinity sign - Analemma or Lemniscates or sideways eight - and the spirals. Both of which, are found throughout the world in many forms. These, we argued, could connect the historical record to space flight. These elements add support, each in however small way, to the conclusion we reached:

The alignments found could not have been achieved by earth bound beings without flight capability, or without the assistance of flight empowered ones.

This research has shown that behind the archaeological record stands a civilization capable of viewing the earth from space. These individuals belonged to an ancient civilization that lived on earth, perhaps as early as 44k years ago.

They measured the earth's axis tilt and left a record of this knowledge. The value of the angle of declination of the earth's axis was enshrined in lines, structures and glyphs at many locations the world over, most of them remote from each other, but connected through their geometries. The entire citadel of Machu Picchu was built to reflect the earth's axis obliquity angle.

The technological acumen of the architects of these monuments is revealed in the encoding of the earth's astronomical properties, at many sites such as Göbekli Tepe, Teotihuacan, Stonehenge, Kalasasaya and the Medicine Wheel, among others. These monuments were astronomical geodetic markers placed at pre determined locations on earth. Their geographical locations and the azimuth alignment of the structures at the site or of a feature in them were designed to encode the direction of the circle of illumination at the given latitude. The sites' longitudes were selected to set the distance between the circles of illumination equally spaced around the Equator which mark the mean solar day divided in quarter hours; in effect that ancient civilization created a precise Global Clock.

Each location has a feature that depicts the declination angle of earth's axis. In some cases the geographic alignment vs. north, if not exact the difference in azimuth points in a predetermined direction. The arquitects used alignments with the Polar star as a tool to indicate the importance of nearby alignments in other directions. We confirmed that many notable structures, ancient or current era, are aligned with the cardinal points. The ones that are not aligned this way, were determined their alignments, invariably, point in a predetermined direction, serving as a guide to reach other specific points, even when found around the world. In the Nasca plain, it is a line that starts at a prominently marked node point, at Machu Picchu the city itself and some of its structures, in Easter Island the direction of Ahu Hanga Poukura, in

Sacsayhuaman the Sun Dial's direction of a radius, in the El Infiernito, one of the megaliths alignment with the central megalith giant, in UK's Stonehenge; specific megaliths and the gates between Trilithons align through a central point, in America's Stonehenge structural alignments and one of the pathways leading to the center, at Avebury, there are two separate radial megalith alignments. Radii from each center parallel to each other follow the angle of the ecliptic and perpendicular to both, a megalith alignment north-south which runs alongside the road has the bearing of the declination angle of the earth's axis tilted away from the sun at the winter solstice. At the Abbas Giant location, the southwest line of the quadrangle closely follows the declination angle of the earth's axis and so does the Atacama Giant's skirt line. This redundant evidence leaves no doubt these important astronomical facts were known to the designers of these locations.

Some of the alignments led to places where mineral resource s are found. The ancients, left a record of all earth has to offer: Uranium, Gold, Copper, Salt, Nitrate, Bauxite, iron and in the Candelabra's design they encoded our atmosphere's gasses: Nitrogen, Oxygen and water vapor.

The foregoing description of the 'stars' can be visualized as the maps we can find on a back page of an airliner's in flight magazine. In those pages there are route maps that are 'stars'. Similarly, the ancients left us a record of the Hubs of their travels and their routes and the structures they built to mark them. We believe the results of this research may dispel the mystery aspect that has been associated with many of these sites as to their purpose. Now we know their purpose: they record travel routes that lead to the location of centers of human presence (hominids?), and or the location of earth's resources. These sites were created perhaps over long time periods or eras.

Although some structures could have been used as calendars, useful as agricultural guidance tools, we can say that in most instances their geographical locations were more important or took precedence over their alignments for agricultural purposes, particularly for those which form part of the Global Clock.

Agriculture flourished around the world in remote places without tools of this nature. Assigning them an agricultural or spiritual function as a rationale for their existence is just romantic, not factual, although those peoples who thrived around these monuments possibly benefited on both counts. The alignments are more appropriately explained as the result of the mapping process. We, now, know that regardless of the location these structures are found in on earth, or the human cultures that thrived around them, all are geometrically aligned, although some of them, perhaps by chance. The physical evidence found thus far, bears this out as anyone can easily prove to themselves following the methodology given in the next chapter.

On this point, there is an ongoing controversy regarding the alignment of the Giza Pyramids with Orion vs. Cygnus and the times these alignments may have occurred in, 10,500 or 2,600 BC; and if one or the other fits better. The results presented by either side of the argument show they do not fit quite exactly for either one of those dates[24]. Based on our results we question whether this argument is material at all. If the starry alignment were to hold factual, regardless of which constellation we choose, then all the alignments we have measured would be dependent on this one alignment, which by its sidereal nature it is a moving target. The Giza Pyramids are far from being one of the most important alignment 'Hubs'. However, the alignments they have, which from our perspective are significant, are the alignments with the South Pacific Ocean ridge line and the 1,400 mile alignment with the South Indian Ocean Plate Ridge. That appears to be a more down to earth alignment anchoring landmark which in turn connects to a global alignment network.

CHAPTER 14

METHODOLOGY
This research was conducted in its entirety with Google$^©$ earth and Wikipedia, Geo-Hack. Wikipedia was used primarily as a site finding tool. The coordinates that are given for sites were confirmed to be quite accurate. Another useful service found there are the categorized archaeological site listings and links.

Google$^©$ earth is easy to use. A location may be searched by name or by coordinates. The names of places, sites, structures, etc. can be found in Wikipedia or are quoted in the text. Once the desired site is found, Wikipedia provides the coordinates to the site. The line drawing and viewing were standardized. The map was correctly aligned north before taking a measurement. The eye altitude was set at about 1,500 ft., doing this provides good reproducibility. Angle readings, in particular, are sensitive to eye height and map viewing direction and location. In most instances the measurements are easy to reproduce to a good approximation. There is a parallax figment of the technology which requires a set viewing angle and position on the screen.

For this work the best way to ensure the direction angle is correct is to draw the great circle for the line, starting at an elevation as close to the ground level as possible. This is not as important for local measurements. The great circle needs to be drawn in two half measurements. At the antipode point where the two half circles meet, the end-point, before 'setting' the line down; click when the desired angle value appears on the pop-up pane. The closer the view is to the ground the more accurately the desired angle can be set.

The circle of illumination for a site or structure is found by extending a line from the object in both directions to the Arctic and Antarctic circles to the tangential point on each side of the circle while zooming in to superimpose the line on the circle. Without setting the line the measurement is recorded and the line continued to the antipode point; its end-point. In the northern hemisphere, the circle to the right of the North Pole is the declination angle at the summer solstice. The circle to the left of the North Pole is the angle at the winter solstice. At the culmination of the summer solstice the sunrise line's azimuth is perpendicular to the winter solstice circle of illumination.

It is useful to label each line with a mind jogging name, length and angle. Although some of these data are automatically collected in the metadata, this way is more expeditious to search for a line or to confirm a piece of data. One can click on a line or marker on the globe to highlight it on the listing, and immediately read off the data. Google[©] earth allows for data sorting only if found within a subfolder. If desired, the data can be separated by creating different myplaces.kml files and saving each folder separately. The myplaces.kml file is found under: C:\Users\username\AppData\LocalLow\ Google\ Google earth. Your folder 'properties' has to be set to 'view hidden files' in order to see the AppData folder.

Kmz files can be ported to Excel[©] via various methods found on the web. The easiest is to save the file as *.xml and opening it with Excel[©]. After accepting 'yes' to a few options the data can be arranged and formatted as desired.

Note: Google©earth .kmz files are available for each topic discussed. Please send your request for a list of available files or information contact at: villarts@charter.net.

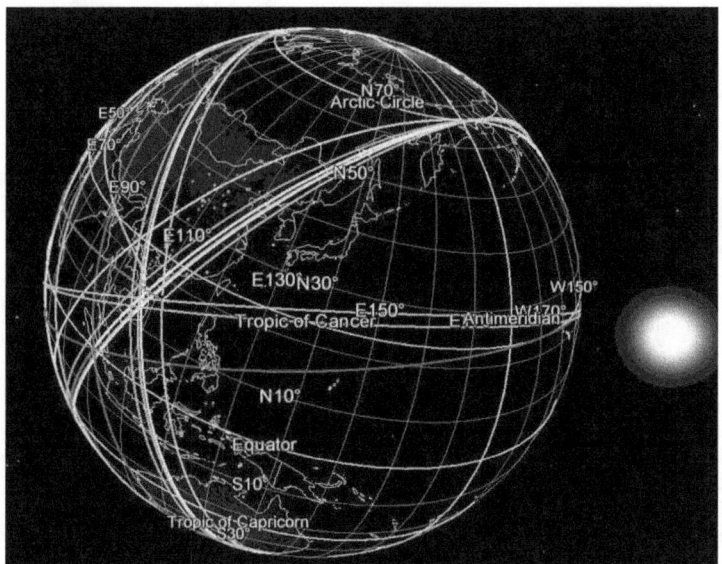

About the Author
The author is a chemist, (SUNY Stony Brook 1976), scientist inventor, holder of several US and foreign patents. The author has traveled extensively throughout the Americas, England and the Far East. His interest in archaeology has been primarily focused on the Americas. He has visited many of the sites mentioned in the book and had the opportunity to observe some of the celestial phenomena; some of the sites are aligned to. This book is the culmination of years of pondering and research resulting in the paradigm changing discoveries documented in this work.

APPENDIX

BIBLIOGRAPHY

1. The International Union of Pure and Applied Chemistry, Zürich, Switzerland.
2. The End of Days, Z. Sitchin. Harper Collins, 2008
3. Pre-Columbian Cities-Jorge E. Hardoy. Walker and Company NY, 1973
4. 1421 The year China Discovered America-Gavin Manzies-William Morrow, 2002
5. Studies on the Population of China, 1368-1953. Harvard University Press; First Edition (US) First Printing edition (January 1, 1959)
6. Mystic Places-Mysteries of the Unknown-Time Life Books, 1987
7. Eye Witness to History. National Geographic Society, 2013
8. Sir Alfred Watkins, Discoverer of Lay Lines; Early British Trackways, 1921, Amazon, Kindle
9. Messagetoeagle.com/aliencarvrockchin.php
10. Lordkipanidze, D. et al. Science 342, 326–331 (2013).
11. Hugh Newman-Earth Grids, Wooden Books, 2008
12. http://home.hiwaay.net/~jalison/- The Prehistoric alignment of World Wonders
13. The Pyramid's alignments-Graham Hancock's Presentation at the 2012 conference. You Tube
14. National Geographic Magazine: 7/2010. The Human Family
15. Sir Norman Lockyer. Stonehenge Decoded 1966-Hawkins, Gerald S. 1963 Nature Publishing Group
16. Jim Marrs. Alien Agenda. You Tube
17. http://www.hellenicaworld.com/Peru/Literature/ Bingham/en/IncaLand.html
18. www.catedraldesal.gov.co/
19. Atlantis, the Antediluvian World, by Ignatius Donnelly, [1882], at sacred-texts.com Ch. V
20. The Glyphs of the Gods, Joseph P. Farell. Adventures Unlimited Press (November 21, 2013)
21. http://www.nps.gov/grca/historyculture/index.htm/ Kennewick Man- F.P. Mc Manamon. 5/2004
22. http://es.wikipedia.org/wiki/Bosque_y_complejo_arqueol%C3%B3gico_El_Cañoncillo
23. http://en.wikipedia.org/wiki/Kolar_Gold_Fields
24. http://Apalachianhistory.net/judaculla rock
25. Ancient Aliens, H2Channel.
26. http://www.bibliotecapleyades.net/piramides/esp_piramide_china_3.htm
27. http://www.viewzone.com/clovisart22.html 'z' Pinch astronomical event.
28. http://www.bibliotecapleyades.net/ciencia/antigravityworldgrid/ciencia_antigravityworldgrid01.htm
29. Joseph, Frank; "The Candelabra of the Andes," The Ancient American, 2:10 no. 10, 1995.
30. kmz. file at villarts@charter.net (originally found http://www.wired.com/wiredscience/2011/06/google-oceans/)
31. http://Carl P. Munk, pyramid matrix.com
32. Skiwatchers. A. Aveni 1997, pp.135–138.
33. http://archyfantasies.com/tag/copper-harbor-petroglyph/Martin 1995
34. NASA.gov
35. Jay Stuart Wakefield, MES & AAPF in http://grahamhancock.com
36. Julius Caesar in his Commentarii de Bello Gallico (50s BCE) Wikipedia
37. http://en.wikipedia.org/wiki/Cueva_de_los_Tayos
38. Video available at: http://www.human-resonance.org /hummingbird _pyramid.html.
39. en.wikipedia.org/wiki/Stonehenge

40. http://whc.unesco.org
41. Sun angle data calculator used; http://susdesign.com/sunangle/index.php
42. https://yaotltecuhtli.wordpress.com/2010/08/09/trecenas/
43. http://erc.europa.eu/succes-stories/modern-human-culture-could-have-emerged-44000-years-ago
44. http://www.mirror.co.uk/news/uk-news/44000-year-old-jawbone-proves-ancestors-of-modern-276597
45. http://yurileveratto.com/articolo.php?Id=82
46. Berger, A.L. (1976). "Obliquity and Precession for the Last 5000000 Years". *Astronomy and Astrophysics* 51: 127–135. Bibcode:1976A&A....51..127B
47. Dictionary of technical terms for aerospace use. NASA Lewis Research Center.
48. Hardoy Jorge E. 1973 Pre-Columbian Cities-. Walker and Company NY
49. http://www.archaeological.org/Archaeology%20of%20America%20and%20Canada.kml.
50. http://www.math.nus.edu.sg/aslaksen/gem-projects/hm/0102-1-pyramids/page1002.htm
51. Z. SITCHIN 2008, The End of Days. Harper Collins
52. http://mw2.google.com/mw-earth-vectordb/blog/Google_Earth_Seafloor Updates_06_2011.kmz.
53. http://www.cheops-pyramide.ch/khufu-pyramid/pyramid-alignment.html
54. DONNELLY IGNATIUS Atlantis, the Antediluvian World, at sacred-texts.com
55. curiosity.discovery.com, www.ancient.eu.com
56. http://www.bbc.co.uk/pressoffice/pressreleases/stories/2002/09_september/13/pyramid_pack.pdf
57. http://archaeology.about.com/od/assyria/fl/The-Aqueduct-at-Jerwan-Iraq.htm
58. http://archaeology.about.com/od/dathroughdeterms/qt/Desert-Kites.htm
59. http://www.antiquity.ac.uk/projgall/beck308/
60. http://www.tandfonline.com/doi/abs/10.1080/00438243.2010.498640#.VACAH6NCJu
61. http://caotigre.blogspot.com/2012/04/mito-la-diosa-bachue.html
62. John N. Miksic Journal of the Malaysian Branch of the Royal Asiatic Society Vol. 59, No. 1 (250) (1986), pp. 27-32- A valley of megaliths in west Sumatra Mahat(Snithger's Aoer Doeri) revisited

INDEX OF ILLUSTARATIONS -1

INDEX OF ILLUSTRATIONS -2

CLOCK TIME DATA 1

ARCHAEOLOGICAL SITES	COORDINATES		CI	TYPE	COUNTRY	DISTANCE	Metric Hours 1603 km			Mean Solar Day 1674.4 km		
	LATITUDE	LONGITUDE	ANGLE			km.	25h	xpecte	Δ	12h	xpecte	Δ
AURDURI MENHIRS	0.0384	100.4803	156.56	MENHIR	INDONESIA	0.00	0.00	0.00	0.00	0.00	0.00	0.00
NAKTA MANDIR	22.0059	82.5722	154.60	TEMPLE	INDIA	870.85	0.54	0.50	0.04	0.52	0.50	0.02
NANDANA SHIVA	23.4500	78.5809	154.30	TEMPLE	INDIA	1231.81	0.77	0.75	0.02	0.74	0.75	0.01
SAS BAHU TEMPLE	24.6769	73.6372	153.04	TEMPLE	INDIA	1662.96	1.04	1.00	0.04	0.99	1.00	0.01
MOENJO DARO	27.3254	68.1334	153.40	CITY	PAKISTAN	2160.42				1.29	1.25	0.04
ACHAEMENID	30.8262	61.6305	152.46	CITY	IRAQ	2429.17				1.45	1.50	0.05
BAYAZEH VILLAGE	33.3338	55.1064	151.58	CITY	IRAN	3207.08	2.00	2.00	0.00	1.92	2.00	0.08
MAYMAND	30.2309	55.3760	152.58	CITY	IRAN	3390.64				2.02	2.00	0.02
NAQSH-E RUTSAM	3.8138	113.7816	152.67	TEMPLE	IRAN	3686.66				2.20	2.25	0.05
CHONGHA ZANBIL	32.0090	48.5217	152.05	CITY	IRAN	4034.45	2.52	2.50	0.02			
QARYAT WASIT ATHARIYAH	32.1832	46.2729	151.96	CITY	IRAQ	4247.07				2.54	2.50	0.04
TELL HALAF	36.8426	40.0701	150.22	TEMPLE	SIRIA	4615.74				2.76	2.75	
GOBEKLI TEPE	37.2232	38.9225	150.02	TEMPLE	TURKEY	4710.00	2.94	3.00	0.06	2.81	2.75	0.06
AL RAJAJEEL CIRCLE	29.8127	40.2196	152.70	CIRCLE	SAUDI	5105.52				3.05	3.00	0.05
ALASID CIRCLE	31.0608	38.1848	153.33	CIRCLE	SAUDI	5247.84	3.27	3.25	0.02	3.24	3.25	0.01
JERASH	32.2775	35.8907	151.93	MENHIR	JORDAN	5419.37						
PETRA	30.3262	35.4424	152.54	CITY	JORDAN	5601.94	3.49	3.50	0.01	3.54	3.50	0.04
MONKWOLSTRUP	54.7181	9.4478	136.47	DOLMEN	DENMARK	5925.16						
MEDICINE WHEEL	44.8265	-107.9210	145.87	CIRCLE	US	6004.67	3.75	3.75	0.00			
BRODGAR RING	59.0015	-3.2298	129.41	CIRCLE	SCOTLAND	6400.16	3.99	4.00	0.01			
LHORA DOLMEN	50.7421	8.6206	141.03	DOLMEN	GERMANY	6660.73				3.98	4.00	0.02
CHURCHILL INUKSHUK	58.7737	-94.1715	50.04	CAIRN	CANADA	6717.72	4.19	4.25	0.06			
APOSTLES CIRCLE	53.9016	-1.8096	137.52	CIRCLE	UK	7324.95	4.57	4.50	0.07			
TARXIEN TEMPLE	35.8692	14.5119	150.59	TEMPLE	MALTA	7535.88				4.50	4.50	0.00
DAMPMESNIL DOLMEN	49.1748	1.6493	142.54	DOLMEN	FRANCE	7651.46	4.77	4.75	0.02			
ABBAS GIANT GLYPH	50.8139	-2.4752	141.00	GLYPH	UK	7885.83				4.71	4.75	0.04
CASTLERUDDERY CIRCLE	52.9912	-6.6367	138.62	CIRCLE	UK	8014.74	5.00	5.00	0.00			

CLOCK TIME DATA 2

ARCHAEOLOGICAL SITES	COORDINATES		CI ANGLE	TYPE	COUNTRY	DISTANCE km	Metric Hours 1603 km/h			Mean Solar Day 1674.4 km		
	LATITUDE	LONGITUDE					25h	xpected	Δ	12h	xpected	Δ
BURREN CIRCLE	53.0391	-9.0937	138.55	CIRCLE	IRELAND	8277.87				4.94	5.00	0.06
Trégunc-SACRIFICIAL TABLE	47.8604	-3.8529	143.63	MENHIR	FRANCE	8426.72				5.03	5.00	0.03
ARLOBI MENHIR	43.0316	-2.8285	147.06	MENHIR	SPAIN	8845.11	5.52	5.50	0.02	5.28	5.25	0.03
C AFRICA MENHIRS	5.8559	15.5926	156.44	MENHIR	C AFRICA	9166.91	5.72	5.75	0.03	5.47	5.50	0.03
MENGA DOLMEN	37.0241	-4.5484	150.16	DOLMEN	SPAIN	9571.22				5.72	5.75	0.03
XERES CROMELEQUE	38.4534	-7.3709	149.46	CIRCLE	PORTUGAL	9763.00	6.09	6.00	0.09	5.83	5.75	0.08
ALTO DA VIGIA NEO VILLAGE	38.8242	-9.4729	149.30	CITY	PORTUGAL	9965.51	6.22	6.25	0.03	5.95	6.00	0.05
VERNEUKPAN	-29.9971	21.1015	332.65	GLYPH	SO AFRICA	10448.09				6.24	6.25	0.01
TIFARITI	26.2017	-10.6031	153.68	GLYPHS	W SAHARA	10991.73				6.56	6.50	0.06
EL RICHAT PICTOGRAPHS	21.1258	-11.4046	154.75	GLYPH	ALGIRES	11380.07	7.10	7.00	0.10	6.80	6.75	0.05
NUNAVUT Inukshuk CAIRN 1	15.7304	-14.0504	155.59	CAIRN	CANADA	11964.58	7.46	7.50	0.04	7.15	7.00	0.15
KIMMIRUT MENHIR	62.8529	-69.8686	119.29	MENHIR	CANADA	12533.80				7.49	7.50	0.01
LAKE MUNGO	-33.7450	143.0813	208.58	BURIAL	AUSTRALIA	13416.52				8.01	8.00	0.01
RIGOLET DIG	54.0745	-58.5767	137.25	CITY	CANADA	13603.52	8.49	8.50	0.01	8.12	8.00	0.12
WHITE ROCK MENHIR	59.4282	-74.7994	128.52	MENHIR	CANADA	14249.86				8.51	8.50	0.01
HUDSON BAY INUKSHUK	58.4682	-78.0945	130.50	CAIRN	CANADA	14870.12	9.28	9.00	0.28	8.88	8.75	0.13
RANKIN INLET INUKSHUK	62.8173	-92.0759	119.44	CAIRN	CANADA	15016.63				8.97	9.00	0.03
KEJIMKUJIK PETROGLYPHS	44.4008	-65.2199	146.14	GLYPH	CANADA	15644.13	9.76	9.75	0.01	9.34	9.25	0.09
AM STONEHENGE	42.8422	-71.2096	147.14	CIRCLE	NH US	16473.59	10.28	10.25	0.03	9.84	9.75	0.09
CHIMENEY POINT	44.0360	-73.4185	146.39	CITY	VT US	16598.23				9.91	10.00	0.09
TIBES CITY	18.0439	-66.6228	335.27	CITY	PR US	17697.70	11.04	11.00	0.04	10.57	10.50	0.07
NEWARK CIRCLES	-82.4311	40.0410	148.68	MOUND	OH US	17980.68	11.22	11.25	0.03	10.74	10.75	0.01
AZTALAN MOUNDS	43.0646	-88.8622	147.00	MOUND	IL US	18415.84	11.49	11.50	0.01	11.00	11.00	0.00
TYRONA CITY	11.0381	-73.9253	156.08	CITY	COLOMBIA	18874.58	11.77	11.75	0.02	11.27	11.25	0.02
EL ABRA TOMB	4.5882	-74.3158	156.49	DOLMEN	COLOMBIA	19237.43	12.00	12.00	0.00	11.49	11.50	0.01
GUDUNG PADANG	6.9930	107.0565	156.37	MENHIR TEMI	INDONESIA	19641.52	12.253	12.25	0.00	11.73	11.75	0.02

CLOCK TIME DATA 3

ARCHAEOLOGICAL SITES	COORDINATES		CI	TYPE	COUNTRY	DISTANCE	Metric Hours 1603 km/			Mean Solar Day 1674.4 km		
	LATITUDE	LONGITUDE	ANGLE			km.	25h	toecte	Δ	12h	toecte	Δ
CHACCHOBEN SUN	19.0012	-88.2304	155.11	PYRAMID	MEXICO	20035.00	12.50	12.50	0.00	11.97	12.00	0.03
CANCUEN PYRAMIDS	15.9975	-90.0441	155.55	PYRAMID	MEXICO	20413.27	12.73	12.75	0.02	12.19	12.25	0.06
WHT MOUNT PETROGLYPHS	41.8911	-109.2653	147.69	GLYPH	US	20800.34	12.98	13.00	0.02	12.42	12.50	0.08
SUN -TEOTIHUACAN	19.6925	-98.8436	155.00	PYRAMID	MEXICO	21190.37	13.22	13.25	0.03	12.66	12.75	0.09
TZINTAUSTZAN PYRAMIDS	19.6241	-101.5739	155.02	PYRAMID	MEXICO	21500.16	13.41	13.50	0.09	12.84	13.00	0.16
SONORA STONES PETROGLYPH	27.8331	-109.4620	153.26	GLYPH	MEXICO	21905.48	13.67	13.75	0.08	13.08	13.00	0.08
CHINA LAKE PETROGLYPHS	35.9939	-117.6117	150.56	GLYPH	US	22234.36	13.87	14.00	0.13	13.28	13.25	0.03
MONGESWAR TEMPLE	10.4638	79.2274	23.87	TEMPLE	INDIA	22918.34	14.30	14.25	0.05	13.69	13.75	0.06
LEPAKSHI TEMPLE	13.8018	77.6094	24.18	TEMPLE	INDIA	23262.92	14.51	14.50	0.01			
PARASGAD	15.7395	75.1290	24.40	TEMPLE	INDIA	23639.84	14.75	14.75	0.00	14.12	14.00	0.12
AL JASSASIYA PETROGLYPHS	25.9516	51.4062	133.72	GLYPHS	QATAR	24146.47				14.42	14.50	0.08
AHU POUKURA	-27.1671	-109.3795	333.43	MENHIR	RAPA NUI	24782.69	15.46	15.50	0.04	14.80	14.75	0.05
ALASID	31.0608	38.1848	332.33	CIRCLE	SOUDI	25285.44				15.10	15.00	0.10
NIKOLSKI MOUND	-168.8750	52.9064	138.74	MOUND	AK US	26089.26	16.28	16.25	0.03	15.58	15.50	0.08
VERNEUKPAN	-29.9971	21.1015	207.34	GLYPH	SO. AFRICA	27257.00	17.00	17.00	0.00	16.28	16.25	0.03
BABYLON	32.5363	44.4209	28.16	CITY	IRAQ	28063.68	17.51	17.50	0.01	16.76	16.75	0.01
NINEVEH	36.3597	43.1516	29.59	CITY	IRAQ	28488.73	17.77	17.75	0.02	17.01	17.00	0.01
EDFU	24.9782	32.8725	26.03	TEMPLE	EGYPT	28857.00	18.00	18.00	0.00	17.23	17.25	0.02
HORANCHOS PICTOGRAPHS	38.5507	-6.0592	329.41	GLYPH	SPAIN	29642.41	18.49	18.50	0.01	17.70	17.75	0.05
SHERE HILLS BAND HOLES	9.9226	8.9851	336.17	GLYPH	NIGERIA	30681.10	19.14	19.00	0.14	18.32	18.25	0.07
TARXIEN TEMPLE	35.8692	14.5119	29.40	TEMPLE	MALTA	31637.12	19.74	19.75	0.01	18.89	19.00	0.11
HA'AMONGA TRILUTHON	-21.1413	-175.0424	334.75	MENHIR	TONGA	31741.59				18.96	19.00	0.04
NUNAVUT Inukshuk CAIRN 1	15.7304	-14.0504	295.95	CAIRN	CANADA	32002.00	19.96	20.00	0.04	19.11	19.25	0.14
LUNAMAROONA	39.6500	8.8795	31.09	DOLMEN	SARDINIA	32618.33	20.35	20.25	0.10	19.48	19.50	0.02
PECH MAHO	43.0458	2.9564	32.99	CITY	FRANCE	33555.00	20.93	21.00	0.07	20.04	20.00	0.04
HOUSE OF TAGA	14.9668	145.6221	155.70	TEMPLE	TINIAN	34303.79				20.49	20.50	0.01
MOCHONG LATTE STONES	14.1984	145.2575	155.79	MENHIR	GUAM	34384.47	21.45	21.50	0.05			
BURYEONGSA temple	36.9407	129.2736	150.14	TEMPLE	SO KOREA	34741.12				20.75	20.75	0.00
HWASUN DOLMEN	34.9714	126.9362	150.98	DOLMEN	SO. KOREA	35162.92	21.94	22.00	0.06	21.00	21.00	0.00
MANROCK CAIRN	45.4799	114.3100	145.46	CAIRN	MONGOLIA	35621.53	22.22	22.25	0.03	21.27	21.25	0.02
HEAVEN TEMPLE	39.8756	116.4069	148.78	TEMPLE	CHINA	35933.64	22.42	22.50	0.08	21.46	21.50	0.04
YONAGUNI PYRAMID	24.4022	123.2946	154.10	PYRAMID	JAPAN	36267.27	22.62	22.75	0.13	21.66	21.75	0.09
TOTEM BAYAN NUR	40.5704	106.3126	148.43	MENHIR	CHINA	36979.25	23.07	23.00	0.07	22.09	22.00	0.09
ASTRONAUT ROCK SILUN	22.6808	111.3383	154.44	MENHIR	CHINA	37697.84	23.52	23.50	0.02	22.51	22.50	0.01
NIAH CAVE 40 KYO	52.9064	-168.8750	156.52	DOLMEN	BRUNEI	38404.73	23.96	24.00	0.04	22.94	23.00	0.06
PRASAT YER TEMPLE	14.8307	104.3702	155.72	TEMPLE	THAILAND	38903.71	24.27	24.25	0.02	23.23	23.25	0.02
PRASAT BAN PRASAT TEMPLE	13.8693	101.7023	155.80	TEMPLE	THAILAND	39248.37	24.48	24.50	0.02	23.44	23.50	0.06
GUNUG PADANG	6.9930	107.0565	336.37	MENHIR TEMI	INDONESIA	39681.31	24.75	24.75	0.00	23.70	23.75	0.05
AURDURI MENHIRS	0.0384	100.4803	156.56	MENHIR	INDONESIA	40075.00	25.00	25.00	0.00	23.93	24.00	0.07

SAJAMA LINE NODES

SAJAMA LINE NODES -1						
	latitude	longitude			latitude	longitude
NODE 01	-18.617215	-68.408924		NODE 29	-18.467077	-68.219774
NODE 02	-18.732446	-68.371000		NODE 30	-18.385580	-68.222441
NODE 03	-18.275435	-68.682263		NODE 31	-18.372748	-68.233665
NODE 04	-18.275157	-68.667510		NODE 32	-18.418738	-68.034133
NODE 05	-18.327598	-68.617383		NODE 33	-18.437782	-68.011751
NODE 06	-18.386235	-68.614488		NODE 34	-18.497132	-68.042735
NODE 07	-18.557204	-68.425253		NODE 35	-18.561639	-68.104804
NODE 09	-18.373106	-68.577454		NODE 36	-18.548193	-68.105015
NODE 10	-18.518111	-68.600172		NODE 37	-18.519346	-68.126266
NODE 11	-18.321770	-68.672596		NODE 38	-18.511159	-68.198184
NODE 12	-18.319840	-68.705882		NODE 39	-18.486492	-68.111738
NODE 13	-18.734067	-68.079472		NODE 40	-18.463369	-68.069379
NODE 14	-18.805536	-67.830661		NODE 41	-18.830597	-67.869668
NODE 15	-18.766191	-67.851533		NODE 42	-18.830597	-67.869668
NODE 16	-18.766191	-67.851533		NODE 43	-18.846862	-67.861437
NODE 17	-18.762640	-67.880744		NODE 44	-18.886499	-67.876432
NODE 18	-18.762640	-67.880744		NODE 45	-18.868917	-67.820098
NODE 19	-18.725512	-67.899414		NODE 46	-18.603289	-68.531927
NODE 20	-18.716575	-67.910029		NODE 47	-18.607550	-68.392378
NODE 21	-18.590998	-68.072353		NODE 48	-18.591459	-68.451807
NODE 22	-18.487644	-68.319485		NODE 49	-18.775306	-67.888634
NODE 23	-18.567475	-68.228254		NODE 50	-18.795567	-67.892716
NODE 24	-18.567475	-68.228254		NODE 51	-18.753241	-67.886441
NODE 25	-18.518488	-68.222219		NODE 52	-18.752458	-67.906781
NODE 26	-18.517328	-68.240168		NODE 53	-18.743233	-67.892007
NODE 27	-18.501309	-68.240073		NODE 54	-18.916460	-68.219774
NODE 28	-18.470145	-68.232427		NODE 55	-18.697860	-68.222441

SAJAMA LINE NODES -2					
	latitude	longitude		latitude	longitude
NODE 56	-18.718007	-67.926381	NODE 84	-18.848264	-67.848383
NODE 57	-18.718635	-68.016639	NODE 85	-18.859274	-67.866240
NODE 58	-18.695284	-68.025270	NODE 86	-18.872449	-67.873666
NODE 59	-18.722937	-68.042072	NODE 87	-18.487126	-68.330501
NODE 60	-18.603141	-68.044760	NODE 88	-18.492734	-68.338590
NODE 61	-18.598220	-68.043283	NODE 89	-18.858621	-68.105395
NODE 62	-18.589797	-68.035520	NODE 90	-18.937261	-67.702862
NODE 63	-18.573696	-68.021609	NODE 91	-18.911005	-67.920041
NODE 64	-18.609941	-68.084135	NODE 92	-18.804050	-67.906669
NODE 65	-18.641144	-67.993592	NODE 93	-18.844472	-67.923971
NODE 66	-18.632077	-67.955515	NODE 94	-18.374562	-68.737567
NODE 67	-18.768930	-67.890243	NODE 95	-18.468149	-68.390927
NODE 68	-18.789586	-67.830538	NODE 96	-18.684604	-68.229810
NODE 69	-18.788419	-67.840726	NODE 97	-18.766844	-68.176602
NODE 71	-18.688343	-67.947270	NODE-101	-18.762259	-68.465345
NODE 72	-18.740728	-67.870356	NODE-102	-18.801641	-68.399840
NODE 73	-18.766766	-67.885401	NODE-103	-18.399122	-68.531245
NODE 74	-18.676943	-67.831378	NODE-104	-18.409300	-68.519817
NODE 75	-18.147996	-68.820222	NODE-111	-18.541921	-67.958025
NODE 76	-18.804408	-67.973656	NODE-105	-18.682376	-68.432493
NODE 77	-18.632847	-67.900233	NODE-106	-18.583688	-68.049940
NODE 78	-18.476268	-68.197727	NODE-106	-18.583688	-68.049940
NODE 79	-18.472961	-68.213348	NODE-106	-18.583688	-68.049940
NODE 80	-19.035786	-67.451601	NODE-107	-18.479471	-68.214533
NODE 81	-18.682144	-67.836338	NODE-108	-18.783257	-67.981244
NODE 82	-18.146425	-68.656593	NODE-109	-18.724091	-68.384794
NODE 83	-18.437048	-68.669057	NODE-109	-18.486125	-68.324826
			NODE-110	-18.573121	-68.024256

SAJAMA KIVAS-Coordinates

	latitude	longitude			latitude	longitude
KIVA	-18.9109	-67.3503		KIVA	-19.1128	-68.0528
KIVA	-18.9188	-67.3545		KIVA	-19.1113	-68.0529
KIVA	-18.9241	-67.3600		KIVA	-19.0775	-68.0532
KIVA	-18.9241	-67.3600		KIVA	-19.0987	-68.0541
KIVA	-18.9260	-67.3626		KIVA	-19.0975	-68.0541
KIVA	-18.9396	-67.7697		KIVA	-19.0500	-68.0541
KIVA	-18.8057	-67.8039		KIVA	-19.0500	-68.0541
KIVA	-18.8088	-67.8076		KIVA	-19.0500	-68.0541
KIVA	-18.8088	-67.8076		KIVA	-19.0497	-68.0542
KIVA	-18.3883	-67.8256		KIVA	-19.0497	-68.0542
KIVA	-18.3883	-67.8256		KIVA	-19.0497	-68.0542
KIVA	-18.3883	-67.8256		KIVA	-19.0497	-68.0542
KIVA	-18.3883	-67.8256		KIVA	-19.0497	-68.0542
KIVA	-18.3883	-67.8256		KIVA	-19.0983	-68.0543
KIVA	-18.3865	-67.8291		KIVA	-19.0983	-68.0543
KIVA	-18.3865	-67.8291		KIVA	-19.0983	-68.0543
KIVA	-18.3865	-67.8291		KIVA	-19.0688	-68.0555
KIVA	-18.3865	-67.8291		KIVA	-19.0688	-68.0555
KIVA	-18.3865	-67.8291		KIVA	-19.0688	-68.0555
KIVA	-18.3865	-67.8291		KIVA	-19.0688	-68.0555
KIVA	-18.3865	-67.8291		KIVA	-19.0695	-68.0560
KIVA	-18.3865	-67.8291		KIVA	-19.0695	-68.0560
KIVA	-18.6801	-67.8351		KIVA	-19.0695	-68.0560
KIVA	-18.3983	-67.8368		KIVA	-19.0695	-68.0560
KIVA	-18.8398	-67.8407		KIVA	-19.0774	-68.0567
KIVA	-18.8412	-67.8421		KIVA	-19.0774	-68.0567
KIVA	-18.8764	-67.8750		KIVA	-19.0774	-68.0567
KIVA	-18.7774	-67.8910		KIVA	-19.0777	-68.0571
KIVA	-18.6328	-67.9002		KIVA	-19.0777	-68.0571
KIVA	-19.0397	-67.9057		KIVA	-19.0777	-68.0571
KIVA	-18.7343	-67.9063		KIVA	-19.0616	-68.0573
KIVA	-18.8041	-67.9067		KIVA	-19.0795	-68.0574
KIVA	-18.7165	-67.9078		KIVA	-19.0771	-68.0574
KIVA	-18.7185	-67.9284		KIVA	-19.0759	-68.0580

	latitude	longitude			latitude	longitude
KIVA	-19.0904	-68.0785		KIVA	-18.9987	-68.0878
KIVA	-19.0904	-68.0785		KIVA	-18.6812	-68.0879
KIVA	-19.0904	-68.0785		KIVA	-19.0806	-68.0880
KIVA	-19.1053	-68.0790		KIVA	-19.0806	-68.0880
KIVA	-19.1053	-68.0790		KIVA	-19.0806	-68.0880
KIVA	-19.0470	-68.0797		KIVA	-19.0800	-68.0883
KIVA	-19.0153	-68.0797		KIVA	-19.0800	-68.0883
KIVA	-19.0153	-68.0797		KIVA	-19.0808	-68.0885
KIVA	-19.1069	-68.0797		KIVA	-19.0854	-68.0886
KIVA	-19.1069	-68.0797		KIVA	-19.0854	-68.0886
KIVA	-19.1069	-68.0797		KIVA	-19.0084	-68.0893
KIVA	-19.1064	-68.0799		KIVA	-19.0845	-68.0893
KIVA	-19.1064	-68.0799		KIVA	-19.0852	-68.0894
KIVA	-19.1064	-68.0799		KIVA	-19.0852	-68.0894
KIVA	-19.0910	-68.0801		KIVA	-19.0852	-68.0894
KIVA	-19.1072	-68.0802		KIVA	-19.0847	-68.0902
KIVA	-19.1072	-68.0802		KIVA	-19.0847	-68.0902
KIVA	-19.1083	-68.0809		KIVA	-19.0847	-68.0902
KIVA	-19.0920	-68.0809		KIVA	-19.0847	-68.0902
KIVA	-19.0889	-68.0810		KIVA	-19.0836	-68.0904
KIVA	-19.0435	-68.0811		KIVA	-19.0846	-68.0911
KIVA	-19.0435	-68.0811		KIVA	-19.0838	-68.0913
KIVA	-19.1132	-68.0812		KIVA	-19.0737	-68.0919
KIVA	-19.0161	-68.0813		KIVA	-19.0737	-68.0919
KIVA	-19.1008	-68.0813		KIVA	-19.0195	-68.0920
KIVA	-19.0174	-68.0815		KIVA	-19.0694	-68.0923
KIVA	-19.0174	-68.0815		KIVA	-19.0241	-68.0923
KIVA	-19.0902	-68.0815		KIVA	-19.0836	-68.0923
KIVA	-19.0766	-68.0819		KIVA	-19.0836	-68.0923
KIVA	-19.0404	-68.0822		KIVA	-19.0836	-68.0923
KIVA	-19.0908	-68.0823		KIVA	-19.0836	-68.0923
KIVA	-19.0993	-68.0824		KIVA	-19.0226	-68.0925
KIVA	-19.0993	-68.0824		KIVA	-19.0226	-68.0925
KIVA	-19.0993	-68.0824		KIVA	-19.0226	-68.0925

CHICHEN ITZA - MEXICO
Photographs by the Author

El Castillo

Ball Court

El Caracol Observatory

www.ingramcontent.com/pod-product-compliance
Lightning Source LLC
Chambersburg PA
CBHW051855170526
45168CB00001B/119